||||||| |||| ||||| ||||| T0214224

 Birkhäuser

Science Networks . Historical Studies
Founded by Erwin Hiebert and Hans Wußing
Volume 41

Edited by Eberhard Knobloch, Helge Kragh and Erhard Scholz

Ad Meskens

Travelling Mathematics - The Fate of Diophantos' Arithmetic

Ad Meskens
Artesis Hogeschool Antwerpen
Department Bedrijfskunde,
Lerarenopleiding en Sociaal Werk
Verschansingstraat 29
2000 Antwerpen
Belgium
e-mail: ad.meskens@artesis.be

Fig. 1.1 © Stephen Chrisomalis, published with kind permission. All drawings by Paul Tytgat, with kind permission.

ISBN 978-3-0348-0314-4   978-3-0346-0643-1 (eBook)
DOI 10.1007/978-3-0346-0643-1

Cover illustration: From Waller Ms de-00215, August Beer: *Über die Correction des Cosinusgesetzes bei der Anwendung des Nicol'schen Prismas in der Photometrie*, after 1850. With friendly permission by The Waller Manuscript Collection (part of the Uppsala University Library Collections).

Cover design: deblik

Printed on acid-free paper

Springer Basel AG is part of Springer Science + Business Media

www.birkhauser-science.com

Contents

Preface

Anyone studying the work of Diophantos of Alexandria is immediately confronted with three questions: 'Who was Diophantos?', 'How large was his algebraic knowledge?' and 'To what extent are Diophantos' writings unique within the classical mathematical corpus?'

Unfortunately, none of these questions can be answered satisfactorily. Not only is there scant biographical evidence to go on, but the Diophantine corpus also remains rather elusive. Depending on one's position, it lends itself to either minimization or *hineininterpretierung*. Nonetheless, Diophantos' writings have, alongside the work of Heron, come to occupy a special position within the Greek mathematical corpus. Or rather within the *known* ancient mathematical corpus. This qualification is not unimportant, particularly as nearly all of our knowledge of ancient mathematics has come to us via copies, implying that it is to an extent filtered by the choices and judgments of mediæval scribes.

Be that as it may, Diophantos' *Arithmetika*, despite its relative isolation within the classical Greek corpus, represents a phase in the evolution from a syntactic to a symbolic algebra, which we may refer to as syncoptic algebra. In the opening chapters, we shall therefore focus first on the development of algebra in Egypt and Babylon before attempting to formulate answers to the aforementioned questions. Subsequently, we shall consider how the *Arithmetika* influenced the work of Pierre de Fermat.

The classic English study in this field is still Sir Thomas Heath's translation of Diophantos' six Greek books. However, since the publication of this seminal work, many new facts have come to light, not in the least through the discovery of some Arabic manuscripts. It would therefore seem worthwhile to provide an updated account of our understanding of Diophantos and his writings.

The idea for this study came to me during my work on sixteenth-century Netherlandish algebra, more in particular the writings of Simon Stevin, who translated four of the six Greek Diophantos books into French. This first and foremost provided the inspiration for a new Dutch translation (the first, in fact, in over 250 years) of the Greek books of the *Arithmetika*. This Dutch version was produced by my wife, a classical scholar. In addition to the translation, however, we felt it necessary also to explain the mathematics. Our edition, published by Antwerp

University College Press (now Artesis University College Antwerp Press) in 2006, therefore has a special layout: the right-hand pages contain the translation, while the left-hand pages provide a mathematical elucidation in modern notation[1]. This is preceded by an introduction, which –due to space restrictions– had to be shortened, leaving us with the feeling that Diophantos and the fate of his books merited a more extensive account. One thing led to another, and hence this monograph.

There are essentially three ways a contemporary author can render ancient mathematics. He can provide a full-length translation of the original text and leave the interpretation to the reader, or he can render the text in modern mathematical notation, or he can use a semi-symbolic notation that some argue can capture the flavour of the ancient text while also offering the conciseness of modern notation. The latter option, however, would leave the reader with such cumbersome formulae as: $\sqrt{} \sqsubset\!\!\sqsupset (\Box a, \Box b) = \sqsubset\!\!\sqsupset (a, b)$, which a contemporary mathematician would immediately translate as $\sqrt{a^2 b^2} = ab$.

We feel that such a semi-symbolic notation fails on both counts: it is too far removed from the original flavour of the text as well as its mathematical content, and consequently there is a risk it will merely alienate the modern reader from the essence of Diophantos. We have therefore chosen instead to cite the original text where appropriate and to indicate which pitfalls may present themselves when reading it. However, as excessive citing would become tedious and distract the modern reader, we also rely on anachronistic mathematical formulae. We feel this approach offers closer insight into the thinking underlying the writings discussed here, while also doing full justice to their ancient authors' accomplishments. Although the contemporary reader must always bear in mind the Italian adage *traduttore traditore* – in respect of both the translation of the ancient text and its mathematical reformulation – we believe that the combination of text and formulae provides a good basis for a better understanding of *the Arithmetika* and related texts, as well as the issues that ancient mathematicians faced in noting down their work.

In the ten-or-so years it has taken to conclude this study, we have been supported by many people and institutions. Over this period, we have contracted debts of gratitude we can hardly contemplate ever being able to repay adequately. We were also fortunate enough to see our Dutch translation of Diophantos being awarded the Jan Gillis prize of the Royal Flemish Academy of Belgium for Science and the Arts, which provided the necessary funds for further research. We are also indebted to Artesis University College for their support through a scientific project grant and for their permission to use parts of our introduction to the Dutch Diophantos translation in this monograph. The present study would never have materialized either without the cooperation of various libraries, particularly the library of the Teacher Training College of the Artesis University College Antwerp, the Erfgoedbibliotheek Hendrik Conscience (Antwerp) and Museum Plantin More-

---

[1]Heath's translation is in fact a compromise between a translation of the Greek text and a retranslation into mathematical symbolism, which has left us with neither a good edition, nor good mathematics. Nonetheless, it succeeds brilliantly in bringing across the Diophantine flavour.

tus (Antwerp). We are also grateful to Stephen Chrisomalis (Wayne State University) for his kind permission to reproduce the illustration of demotic figures. Special thanks are due to two colleagues for reading drafts of the manuscript and for making useful suggestions: Guido Vanden Berghe and, as always, Jean Paul van Bendegem. The long-standing cooperation with Paul Tytgat (Department of Industrial Sciences) once again proved invaluable: most of the drawings in this book are by his hand. Stephen Windross read and corrected the English draft.

Finally, I must thank my wife Nicole for her continuous support over the past twenty years. Her contribution was a genuine labour of love.

Ad Meskens, 2010

Chapter 1

Arithmetic and the beginnings of algebra

1.1   In the beginning. . .

The origins of arithmetic and algebra are shrouded in the mists of time. It is well known that even the simplest problem in arithmetic can give rise to expressions that we would refer to as equations. All early societies for which we have a written record show evidence of such algebraic problems and their solutions.

The Babylonian and Egyptian civilizations, which have some bearing on the issue at hand, are no exceptions. In Babylonian and Egyptian texts alike, we encounter simple mathematical problems, often relating to the division of possessions. This evidence, like that from many other cultures, seems to confirm that a centrally governed state relies on two prerequisites: grain to feed the people and mathematics to distribute it fairly[1].

Few documents from ancient Egypt –or Greece for that matter– have survived, mainly because the Egyptians wrote on papyrus. To make a papyrus roll, the stem of the papyrus plant (*Cyperus papyrus*) was stripped of its outer rind. The remaining sticky fibrous inner pith was then cut into small pieces measuring about 40 cm, which were subsequently split lengthwise. The papyrus strips were soaked in water and then placed alongside one other, overlapping slightly, so that they could be glued together with a thin floury paste. Another layer of strips was laid transversely. The mould was then put under a press for a couple of days. The separate pieces were glued together into strips of about 10 m long and 20 to 30 cm high.

---

[1] See J. HØYRUP(1994), pp.45-88, R.K. ENGLUND(2001), D.J. MELVILLE(2002), A. IM-HAUSEN(2003), pp.93ff.

A. Meskens, *Travelling Mathematics - The Fate of Diophantos' Arithmetic*, Science Networks.
Historical Studies 41, DOI 10.1007/978-3-0346-0643-1_1, © Springer Basel AG 2010

In a dry climate like that of Egypt, papyrus is stable, formed as it is of highly rot-resistant cellulose; storage in humid conditions however can result in moulds attacking and eventually destroying the material[2]. However, even in favourable climatic conditions, Egyptian papyrus remained vulnerable to other dangers, including rodents.

Three distinct kinds of script were used in ancient Egypt: hieroglyphic, hieratic and demotic. The earliest of the three, hieroglyphic script, was developed as early as 3250 B.C. Hieratic notation emerged in the twenty-sixth century B.C. as a shorthand for hieroglyphic. It developed into many regional variants and, by the late eight century B.C., writing in the Nile Delta had diverged from that in the Upper Nile. From this northern variant of hieratic script developed demotic script, which became the prevailing notation in the Late and Ptolemaic periods and would survive up to around 450 A.D.[3]

Each of the Egyptian scripts had its own system for writing numerals. Demotic numeric script diverges structurally from its hieratic and hieroglyphic counterparts. The latter two are structurally similar to Roman numerical system, except that they were strictly decimal, i.e. without a quinary sub-base[4]. The demotic numeric system on the other hand had a decimal base, with signs for each multiple of each positive integer power of ten, and it was written from right to left[5]. We can therefore see this positional system as an intermediate step between an additive notation and a decimal position system.

	1	2	3	4	5	6	7	8	9
1s									
10s									
100s									
1000s					N/A		N/A		

1/2	1/3	2/3	1/4	1/6	5/6

Figure 1.1 *Demotic numerals (from S. Chrisomalis (2003))*

---

[2]About the making of papyrus and its preservation see R. PARKINSON & S. QUIRKE(1995) and W.E.H. COCKLE(1983).

[3]A. IMHAUSEN(2003), p.3, S. CHRISOMALIS (2003) & (2010), pp.54-56.

[4]On hieroglyphic and hieratic numerals, see L.N.H. BUNT et al.(1976), pp.1-41, G. ROBINS & C. SHUTE(1987), M. KLINE(1972), pp.15-18.

[5]S. CHRISOMALIS(2003),(2004) & (2010), On demotic mathematical papyri, see R.A. PARKER(1972).

For example: ⌐⌐ = 24, ⊣⊣⊣ = 452, ⌐⊣ = 305.

Although the system is hardly ever mentioned in histories of mathematics, it may well have had a significant impact on numerical notations in the Eastern Mediterranean region[6], including Greece, as will become clear in our discussion of Ionian numerals.

The nature of Egyptian mathematical problems may be characterized as numerical, rhetorical and algorithmical. They are *numerical* because they invariably use concrete numbers. They are *rhetorical* because no symbolism is used in formulating the operations. And they are *algorithmical* because the solution is formulated as a sequence of instructions[7].

Egyptian mathematical papyri are either table texts or problem texts. Table texts consist of mathematical data, such as fractions or square roots, arranged in lists. These data are used for solving mathematical problems. Problem texts, on the other hand, put forward mathematical exercises, with or without a practical background or application, and their solutions.

The small body of surviving Egyptian documents with mathematical texts can be divided into three distinct groups[8].

The first group consists of two papyri and some scattered fragments[9]. These documents are written in hieratic script, dating back to the earlier part of the second millennium B.C. The second and third groups date back to the Hellenistic and Roman periods and are in demotic and Greek respectively.

The most complete Egyptian document in the first group of mathematical texts is the Rhind Mathematical Papyrus (also known as the Ahmes Papyrus), which dates from the middle of the sixteenth century B.C.[10] The text contains eighty-four mathematical problems and it is a typical recombination text[11]. While it is mathematical in content, it was probably intended as an instruction manual for administrators.

The other papyrus is Papyrus Moscow, which was discovered at an unknown location. It contains twenty-five mathematical problems[12].

---

[6]On this subject, see S. CHRISOMALIS(2003), (2004) and his seminal study (2010).

[7]A. IMHAUSEN(2002), p.149.

[8]J. FRIBERG(2006).

[9]L.N.H. BUNT et al.(1976), pp.1-41, G. ROBINS & C. SHUTE(1987), A.M. WILSON(1995), pp.19-52.

[10]The document was named after Alexander Henry Rhind, who purchased it in 1858 in Thebes. Rhind, a Scottish scholar, lived in Egypt for health reasons. After his death in 1863, the papyrus was acquired by the British Museum. The copyist of the papyrus identifies himself as Ahmose and mentions that he is writing in the fourth month of the inundation season in the year 33 of the reign of King Ausenes (Apophis). This would put him in the middle of the sixteenth century B.C. See G. ROBINS & C. SHUTE(1987), esp. p.11, for the history of the text.

[11]*Recombination text* is a term coined by J. Friberg to indicate texts containing exercises compiled from other texts. They were arranged in collections of comparable exercises.

[12]The Moscow Mathematical Papyrus is also known the Golenischev Mathematical Papyrus,

Nearly all Old Egyptian problems are linear, solved with the method of duplication and halving. All quadratic equations are of the type $ax^2 = b$.
Multiplication was done by repeated duplications and by addition of the relevant products.

For instance[13]:

12 x 12 (part of problem 32 from Rhind)

1	12
2	24
/4	48
/8	96

Now $12 = 4 + 8$, so

$$
\begin{aligned}
12 \text{ x } 12 &= 48 + 96 \\
&= 144
\end{aligned}
$$

A slash was used to indicate terms that were to be added.

Fractions were calculated by using unit fractions[14], and written as, say,

$$\frac{7}{29} = \frac{1}{6} + \frac{1}{24} + \frac{1}{58} + \frac{1}{87} + \frac{1}{232} = \overline{6} + \overline{24} + \overline{58} + \overline{87} + \overline{232}$$

By way of example, we give the solution method for a simple problem from Rhind papyrus (problem 24)[15]:

An amount added to its one-seventh equals 19. What is the amount?

Suppose $x = 7$,
we find $7 + 7.\overline{7} = 8$
So 8 has to be multiplied as many times to give 19, which multiplied by 7 gives the desired result.

It is found that 8 has to be multiplied by $\dfrac{19}{8}$

$$
\begin{aligned}
\left[\frac{19}{8}\right] &= 2 + \overline{4} + \overline{8} \\
7.\left(2 + \overline{4} + \overline{8}\right) &= 16 + \overline{2} + \overline{8} \\
&\left[= 16\tfrac{5}{8}\right]
\end{aligned}
$$

after its first owner, Egyptologist Wladimir Golenischev. It later entered the collection of the Pushkin State Museum of Fine Arts in Moscow, where it has remained to this day. It probably dates back to the Eleventh Dynasty of Egypt (ca. 2050-1990 B.C.). See V.V. STRUVE & B. TURAEV(1930).

[13]S. COUCHOUD(1993), p.108, G. ROBINS & C. SHUTE(1987), p. 39.

[14]The use of unit fractions in calculations was a commonly applied technique in both Greece and Rome. See also D.W. MAHER & J.F. MAKOWSKI(2001).

[15]G. ROBINS & C. SHUTE(1987), pp. 37-38, S. COUCHOUD(1993), pp.113-114, A. IMHAUSEN(2003b), pp.41 and 206.

Test:

$$16 + \overline{2} + \overline{8} + \overline{7}\left(16 + \overline{2} + \overline{8}\right) = 19$$

The problem is essentially a problem of the type $ax = b$ and hence it can be solved on the basis of proportionality. To this end, the Egyptians used the rule of false value. If a given value does not yield the solution, proportionality dictates that the real value is easily calculated using the rule of three. In this example, 7 is inserted in the first step in place of the unknown, which yields 8. Since the desired number is 19, 8 is multiplied by $\frac{19}{8}$. Consequently, the initially proposed value must also be multiplied by this number, which gives the solution $16\frac{5}{8}$.

Papyri Rhind and Moscow suggest that Egyptian mathematics was not well developed. However, it would be unfair to judge ancient Egyptian mathematics on the basis of just a few very ancient sources.
For instance, in the fragmentary Papyrus Berlin[16], dating back to the Middle Kingdom and written in hieratic script, we encounter the equivalent to systems of equations of the type $\begin{cases} x^2 + y^2 &=& a^2 \\ y &=& bx \end{cases}$.
Obviously, the solution was not given in a formal language, but rather in the shape of an example. The formulation of the problem is interesting, because it is an early predecessor to a Diophantine problem (II.8). Problem 1 of Papyrus Berlin reads[17]:

Two quantities are given. One is $\frac{1}{2} + \frac{1}{4}$ of the other.
The sum of the squares with these quantities as sides is 100.
What are the quantities?

Take a square with 1 as its side. Then the other square has $\frac{1}{2} + \frac{1}{4}$ as its side.
The area of the first square is 1, and the area of the second square is $\frac{1}{2} + \frac{1}{16}$.
The sum of the areas is $1 + \frac{1}{2} + \frac{1}{16}$.
The square root of this sum is $1 + \frac{1}{4}$ and the square root of 100 is 10.
Divide 10 by $1 + \frac{1}{4}$.

---

[16]Papyrus Berlin 6619 is an ancient Egyptian papyrus document from the 19th dynasty (13th century B.C.). This papyrus was found at the Saqqara burial ground in the early 19th century. It contains ancient Egyptian mathematical and medical knowledge, including the first known documentation concerning pregnancy test procedures. See en.wikipedia.org/wiki/Berlin_papyrus
[17]See S. COUCHOUD(1993), pp.131-134, A. IMHAUSEN(2003b), pp.49-50, 53 and 359.

This yields 8, which is the first quantity.

Multiply 8 by $\frac{1}{2} + \frac{1}{4}$, which yields 6, the other quantity.

Again, the solution relies on the rule of false value. The choice of the data is interesting, because $8^2 + 6^2 = 10^2$. So what the Egyptians were effectively doing here was dividing a given square into two other squares. Clearly the compiler of the exercise must have known that there are special triplets for which Pythagoras' theorem holds.

The second exercise of Papyrus Berlin equals the sum of two squares, the second of which is three-quarters of the first, to 400. The solution is of course 16 and 12 $((2.8)^2 + (2.6)^2 = (2.10)^2)$. In fact, apart from the factor, the exercise is identical to the previous one.

The second group of mathematical manuscripts, written in demotic script, show a mathematical evolution that is in line with that observed in comparable other societies.

The most important, and possibly oldest, document of this group is Papyrus Cairo, dating back to the third century B.C.[18]

Typical problems in this papyrus are pole-against-the-wall and reshaping-rectangle problems. In the first kind of problem, the top of a pole of length d, leaning against a wall, slides down a distance p, so that its base moves out from the wall a corresponding distance s. Two of the parameters d, h, s are known, the third needs to be found. Obviously, the solution requires the use of Pythagoras' theorem. Problems such as these have a long tradition and appear in various guises in many different cultures[19].

In reshaping-rectangle problems, the question is: if the width (or height) of a rectangle is decreased by a fraction $\frac{1}{n}$, then by which fraction should the height (resp. width) be increased in order to keep the area constant?

Papyri such as this one may have had a considerable impact on early Greek mathematics. After all, this was the period when Euclid worked in Alexandria.

Whereas our knowledge of Egyptian mathematics is rather limited, we know quite a lot about Babylonian mathematics, because the Babylonians wrote on clay tablets[20]. Clay tablets, unlike papyrus, are not biodegradable and therefore more likely to withstand the passing of time.

Although we speak of *Babylonian* mathematics –in reference to Babylon, one of the most influential cities in the Middle East– the term is somewhat misleading.

---

[18]P. Cairo J.E. 89127-30, 8913. It has a legal code on the recto and some forty mathematical problems on the verso. It was dated on paleological grounds from a study of the legal text. See D. MELVILLE(2004), p.155, J. FRIBERG (2006), pp.105-106.

[19]For a detailed analysis of this type of problem in Egyptian and Babylonian mathematics, see D.J. MELVILLE(2004).

[20]The seminal text on the subject is J. HØYRUP(2002).

Unlike Egypt, Mesopotamia was not a stable culture. The oldest cultures in Mesopotamia, the region between the Tigris and the Euphrates, were those of the Sumerians and the Akkadians. Their reign is referred to as the Old Babylonian period (before 1600 B.C.), an era that is relatively well documented in cuneiform writings. Then came the Hittites, Assyrians, Chaldeans, Medes and Persians. Sometimes these cultural transitions were peaceful, resulting in a more or less continuous evolution of mathematics among other things. At other times they were violent, resulting in breaks in the written tradition. There are, for example, very few written sources from the Kassite era (ca. 1600-1200 B.C.). Moreover, from the twelfth century onward, new technologies for writing were introduced: ink on perishable materials for Aramean and wax-covered wooden tablets for cuneiform writing. Consequently, as in the case of ancient Egypt, few written records from this period have survived.

In 330 B.C., the Middle East was conquered by Alexander the Great. Although his empire was short-lived, it was the beginning of an era of Greek domination that would last until the Arab conquests in the seventh century.

The Sumerians had developed an abstract form of writing based on cuneiform symbols. These symbols were imprinted with a stamp in wet clay, which would subsequently be baked in the sun.

Of the 500000 such clay tablets that have been excavated, around 500 are of mathematical interest. Some 160 tablets containing problems have been published thus far, the vast majority of which date back to the Old Babylonian era[21]. Some tablets contain information that is of indirect relevance to the history of mathematics. For instance, some Old Babylonian mathematical texts deal with various quantities in the context of the digging of a canal[22]. Interestingly, in the past decade or so, a number of school tablets have been described, including some with mathematical content[23]. These tablets give us an insight into the education of administrators. Metrology ran right through the curriculum, beginning with memorization of ordered lists of metrologically related objects up to contextualized metrology in model contracts. Calculations belonged to the advanced curriculum. Mathematical problems were taken from "textbooks" –now referred to as problem texts– often containing quite similar questions with numerical answers.

Most of these mathematical tablets –like the Egyptian papyri– are either table texts or problem texts. Although the tablets usually date back to the Hammurapi Dynasty (18th-16th centuries B.C.) and the Seleucid period (3rd-2nd centuries B.C.), the oldest originated in the twentieth century B.C. It is from these documents that we know the Babylonians were already able to solve quadratic equations around 2000 B.C.[24]

---

[21]E. ROBSON(1999), pp.7-8.
[22]See K. MUROI(1992), J. FRIBERG(2003).
[23]E. ROBSON(2002).
[24]I. BASHMAKOVA & G. SMIRNOVA(2000), pp.1-2.

Unlike the Egyptian texts, which contain routine problems with no apparent connection, the Babylonian texts often present us with carefully arranged problems in increasing order of difficulty[25]. Babylonian mathematics originated and evolved in a practical setting: the orally based surveyor's algebra and the bureaucratic culture of accounting[26]. The legacy of this origin is that the unknown is rendered with a context word, such as *length* (x), *width* (y) or *area* (xy).

Linear equations and systems of linear equations represented no great challenge to the Babylonians. Although the texts only show the calculations, it is not difficult to discern the underlying algorithm of the combination method[27].

Babylonian	*Modern*
$7 \times 4 = 28$	$\begin{cases} l + \dfrac{b}{4} = 7 \\ l + b = 10 \end{cases}$
	$\Rightarrow \begin{cases} 4l + b = 28 \\ l + b = 10 \end{cases}$
$28 - 10 = 18$	$3l = 18$
$18 \times \dfrac{1}{3} = 6$ length	$l = 6$
$10 - 6 = 4$ width	$b = 10 - 6 = 4$

Square roots are found in the calculations of diagonals of rectangles, the so-called *square-side rule*[28]. More often than not, the answer is given without an explanation, but we can deduce from these solutions that the formula relied upon was :

$$d = l + \frac{b^2}{2l}$$

in which d stands for the diagonal, l for the length and b for the width of the rectangle. The result is of course an approximation of the square root, which gets more accurate if b is much smaller than l[29].

---

[25]This may be a little unfair on the Egyptians, as we are comparing a large set of Babylonian tablets with about half a dozen Egyptian papyri.

[26]E. ROBSON(2001), p.170.

[27]L.N.H. BUNT et al.(1976), pp.51-52.

[28]J. FRIBERG(2006), p.82. The rule is of course equivalent to Pythagoras' theorem and states that the square on the diagonal is equal to the sum of the squares on the length and the width of the rectangle.

[29]This has a mathematical basis:

if $b << l$ then $\sqrt{l^2 + b^2} = \sqrt{l^2 \left(1 + \dfrac{b^2}{l^2}\right)} \approx l \left(1 + \dfrac{b^2}{2l^2}\right) = l + \dfrac{b^2}{2l}$.

For a detailed explanation of Babylonian square-root extraction, see D. FOWLER & E. ROBSON(1998), pp.370-373.

Babylonian texts merely provide numerical examples. Nevertheless, they serve as models for a solution method for a particular type of equation. The underlying algorithms are easily discerned. The Babylonians knew that quadratic equations had two distinct, but basically equivalent, solutions. Negative solutions, however, were not considered.

For example[30]

Babylonian algorithm	Modern
Length plus width is 14. Area is 45. What are the length and the width?	*Find two numbers of which the sum is 14 and the product is 45.*
Take half the sum of the length and width (the half-sum): 7	Two numbers that have a sum 14 can be written as $7+x$ and $7-x$, from which
	$(7-x)(7+x) = 45$
Square the half-sum: 49	$\Leftrightarrow 49 - x^2 = 45$
Subtract the area: 4	$\Leftrightarrow x^2 = 4$
Take the square root: 2	$\Leftrightarrow x = 2$
Length is	The numbers are 5 and 9.
half-sum + square root: 9	
Width is	
half-sum - square root: 5	

Systems of equations leading to quadratic equations can be divided into two groups[31]:

$$\begin{cases} x \pm y &= a \\ xy &= b \end{cases} \qquad \begin{cases} x \pm y &= a \\ x^2 + y^2 &= b \end{cases}$$

Quite often, problems are posed in which a number is sought whose sum with its reciprocal is a given number.

This leads to a system of equations: $\begin{cases} xy &= 1 \\ x+y &= b \end{cases}$ which of course can be solved by a quadratic equation:

$$x(b-x) = 1 \Leftrightarrow x^2 - bx + 1 = 0$$

The solution of this equation is given step by step:

calculate $\left(\dfrac{b}{2}\right)^2$, then $\sqrt{\left(\dfrac{b}{2}\right)^2 - 1}$, leading to the solution $\dfrac{b}{2} \pm \sqrt{\left(\dfrac{b}{2}\right)^2 - 1}$.

---

[30]L.N.H. BUNT et al.(1976), p.52. Diophantos' I.27 is the same problem, but with indeterminate sum and product.

[31]K. VOGEL(2004), pp.226-227.

One of the ways of reaching a general solution for the quadratic equation $t^2 + bt + a = 0$ is by writing it as a system[32]:

Given the system $\begin{cases} xy & = & a \\ x+y & = & b \end{cases}$

Then $\dfrac{x+y}{2} = \dfrac{b}{2} \Rightarrow \left(\dfrac{x+y}{2}\right)^2 = \left(\dfrac{b}{2}\right)^2 \Rightarrow \left(\dfrac{x+y}{2}\right)^2 - xy = \left(\dfrac{b}{2}\right)^2 - a$

Now $\left(\dfrac{x+y}{2}\right)^2 - xy = \left(\dfrac{x-y}{2}\right)^2.$

Whence $\left(\dfrac{x-y}{2}\right)^2 = \left(\dfrac{b}{2}\right)^2 - a$

and $\dfrac{x-y}{2} = \sqrt{\left(\dfrac{b}{2}\right)^2 - a}.$

Now $\begin{cases} x = \dfrac{x+y}{2} + \dfrac{x-y}{2} & = & \dfrac{b}{2} + \sqrt{\left(\dfrac{b}{2}\right)^2 - a} \\[2ex] y = \dfrac{x+y}{2} - \dfrac{x-y}{2} & = & \dfrac{b}{2} - \sqrt{\left(\dfrac{b}{2}\right)^2 - a} \end{cases}$

As in Egyptian texts, we find problems relating to the division of a square into two other squares, for which the ratio is given. And, like in the Egyptian sources, we encounter pole-and-reed problems, to be solved on the basis of Pythagoras' theorem.

More importantly, however, there are also purely mathematical exercises, without any immediately apparent practical merit. Among the mathematical texts from Susa is a document designated TMS1, the first problem in which seems to ask for the radius of a circumscribed circle to a given triangle[33]. The height and the front of the triangle are respectively $h = 40$ and $s = 60$. Because the triangle is equilateral, it consists of two right-angled triangles of the (3, 4, 5) type. The values for r (the radius of the circumscribed circle) and of q (the distance from the base to the centre of the circle) are solutions for the system of equations:

$$\begin{cases} r^2 - q^2 & = & \left(\dfrac{s}{2}\right)^2 \\ r+q & = & h \end{cases}$$

---

[32] A procedure also followed by Diophantos in I.27.

[33] J. HØYRUP(2002), pp.265-268, J. FRIBERG(2006), pp.138-139, P. DAMEROW(2001), pp.287-288.

The first equation can also be written as $(r - q)(r + q) = \left(\dfrac{s}{2}\right)^2$ and substituting the second equation, we find $(r - q)h = \left(\dfrac{s}{2}\right)^2$.

The system then becomes:

$$\begin{cases} r - q &= \dfrac{\left(\dfrac{s}{2}\right)^2}{h} = p \\ r + q &= h \end{cases} \Rightarrow \begin{cases} r &= \dfrac{h + p}{2} \\ q &= \dfrac{h - p}{2} \end{cases}$$

The tablet known as Plimpton 322, which has been dated back to between 1900 and 1600 B.C., features a table with numbers that seem to relate to Pythagorean triplets. The tablet consists of four columns with 15 rows of data written in a mixture of Sumerian and Akkadian. The two central columns are entitled *the square of the diagonal* and *the square of the short side*, although the figures are the lengths of the sides of these squares[34].

Tablet IM 67118 contains another predecessor of Diophantine problems[35]. The age of the tablet is known, as the scribe mentions he wrote it during the reign of Ibalpiel II of Eshnunna (fl. 1780-1760 B.C.). It contains a problem that asks for the sides of a rectangle with diagonal 1;15 ($= 75$)[36] and area 45. The solution is provided in a characteristic way: calculation and intermediate results are given until finally the solution, 45 and 1;00 ($= 60$), is arrived at. The problem can be written algebraically as

$$\begin{cases} x^2 + y^2 &= a^2 \\ xy &= b \end{cases}$$

Evidently Pythagoras' theorem was known to the Babylonians, at least at an algorithmic level[37].

The Babylonians even succeeded in solving certain types of cubic equations. They constructed tables for $n^3 + n^2$, with the aid of which equations of the type $ax^3 + bx^2 = c$ can be solved[38].

---

[34] For a recent discussion and interpretation of Plimpton 322, see E. ROBSON(2001).

[35] In book 6, Diophantos deals with right-angled triangles with sides in rational numbers and satisfying conditions. In problems 3 and 4, the additional condition is $\dfrac{1}{2}ab \pm m = a^2$, where a and b are the lengths of the perpendiculars and m is an arbitrary number. Hence $m = 0$ is also solved, which is a restricted version of the problem posed here.

[36] Babylonian figures are sexagesimal, numbers before the semicolon separates are multiples of sixty. Thus $1;15 = 1*60 + 15 = 75$.

[37] For a detailed analysis of the evidence on Pythagoras' theorem in Mesopotamia, see P. DAMEROW(2001).

[38] J.J. JOSEPH(1992), p.207, J. HØYRUP(2002), p.149-154.

Indeed:

$$ax^3 + bx^2 = c$$
$$\Leftrightarrow \left(\frac{ax}{b}\right)^3 + \left(\frac{ax}{b}\right)^2 = \frac{ca^2}{b^3}$$

Put $t = \dfrac{ax}{b}$ then $t^3 + t^2 = \dfrac{ca^2}{b^3}$.

By looking up the value n for which $n^3 + n^2$ equals $\dfrac{ca^2}{b^3}$ the solution for x can be calculated.

On the inheritance from these two cultures, Classical Greece would erect a mathematical monument that continues to casts its shadow forward to this day, namely deductive mathematics.

1.2 Classical Greece

When we speak of Greece, we tend to think of the Peloponnesian mainland and the myriad of islands scattered across the Ionian Sea. In classical times, however, the Greek sphere of influence was often much broader. Greek culture spread out to Southern Italy and Sicily in the west, to Asia Minor in the east, and to the Nile Delta in the south. It reached its zenith at the time of Alexander the Great, whose empire included the whole of the Middle East and parts of India and stretched out as far as Afghanistan.

In 332-331 B.C., Alexander founded a city that still bears his name: Alexandria[39], the ancient centre of learning. Up to the Arab conquests of the seventh century, this city was home to some of the most remarkable mathematicians of the era – including Diophantos.

After Alexander's death, his empire became the scene of a power struggle between his generals. Some tried to establish their own realms in one of the many Persian satraps, and the most ambitious even attempted to seize power throughout the Empire. Eventually, the Greek mainland became an arena for fighting between the city-states and Macedonia.

Ptolemy secured his position in Egypt and became the founder of the Ptolemaic dynasty. Greeks and Macedonians formed the ruling class and the backbone of the army and the bureaucracy. Under their rule, Alexandria became the most important trade hub in the known world. It was also under the Ptolemies that a university and a museum with a library were established, making Alexandria the centre of scientific activity in Ancient society.

Seleucus captured the territory stretching out from the Aegean to India, including the coveted prizes of Syria and Mesopotamia. Seleucus and his successors would establish numerous Macedonian and Greek settlements. New settlers soon mixed with the indigenous populations. The ruling classes and their cities became Hellenized, while the countryside remained Aramean. Hellenization was a deliberate

---

[39]On the foundation of Alexandria: R. CAVENAILLE(1972), M. WOOD(2001), pp.82-83.

policy on the part of the Seleucid dynasty, and it proved particularly successful in Syria. Antioch became the largest Greek city after Alexandria.

Asia Minor remained the centre of Greek culture. The old cities on the Ionian coast grew larger, wealthier and more powerful than ever before. Pergamum developed into a centre of learning to rival Alexandria.

In a sense, then, from 300 B.C. onwards, the mathematics of both Egypt and the Seleucid Empire were "Greek". In this book, however, we shall, in line with common practice, limit the term "Greek mathematics" to the mathematics that originated in the Greek sphere of influence in the Mediterranean.

It is clear that Greek mathematics was influenced by Babylonian and Egyptian sources. To what extent this was the case is, however, a different matter. Herodotos, for example, writes[40]:

> For this reason Egypt was intersected. This king also (they said) divided the country among all the Egyptians by giving each an equal parcel of land, and made this his source of revenue, assessing the payment of a yearly tax. And any man who was robbed by the river of part of his land could come to Sesostris and declare what had happened; then the king would send men to look into it and calculate the part by which the land was diminished, so that thereafter it should pay in proportion to the tax originally imposed. From this, in my opinion, the Greeks learned the art of measuring land; the sun clock and the sundial, and the twelve divisions of the day, came to Hellas from Babylonia and not from Egypt.[41]

Thales[42], Pythagoras, Demokritos and many other Greek mathematicians from the pre-Alexandrian era are all believed to have travelled to Egypt to learn their art. According to Plato, mathematics, which he understood to encompass arithmetic, logistic and astronomy, originated in Egypt in the vicinity of the Greek colony of Naukratis[43].

---

[40] Herodotos, *The Histories*, II.109.

[41] A similar picture is painted by Proklos in his *Commentary on Euclid's* Elements, *Book I*: "[...] it was, we say, among the Egyptians that geometry is generally held to have been discovered. It owed its discovery to the practice of land measurement. For the Egyptians had to perform such measurements because the overflow of the Nile would cause the boundary of each person's land to disappear. [...] And so, just as the accurate knowledge of numbers originated with the Phoenicians through their commerce and their business transactions, so geometry was discovered by the Egyptians for the reason we have indicated." See http://www-history.mcs.st-andrews.ac.uk/Extras/Proclus_history_geometry.html

[42] "Geometry was originally invented by the Egyptians, it was brought to the Greeks by Thales", Heron, *Definitions*, 136,1; "It was Thales, who, after a visit to Egypt, first brought this study to Greece. Not only did he make numerous discoveries himself, but laid the foundations for many other discoveries on the part of his successors, attacking some problems with greater generality and others more empirically.", Proklos, *Commentary on Euclid's* Elements, Book I. See http://www-history.mcs.st-andrews.ac.uk/Extras/Proclus_history_geometry.html

[43] Plato, *Phaedrus* 274c-274d:

Whether there was indeed ever such a direct influence from Babylonia is questionable. The first Greek philosopher to visit Babylon would appear to have been Demokritos, after 449 B.C. This was during a peaceful interlude in the series of wars between Greece and Persia. It was undoubtedly through these Graeco-Persian wars that the Greeks learnt about Persian engineering and associated mathematics and vice versa. On the other hand, there is no evidence for a direct Babylonian influence on Greece insofar as concerns "pure" mathematics[44]. Some writers do however suggest that such an influence existed between Babylon and Egypt, and that the Greeks became acquainted with Babylonian mathematics via this detour, without ever acknowledging this[45].

It has become almost proverbial that Greece is the cradle of science and pure mathematics, with geometry and Euclid's *Elements* in particular as its pinnacle. However, the question arises whether the Greeks themselves also held this view. In fact, they almost certainly did not, as they regarded arithmetic to be superior to geometry[46]. Of course, Greek mathematics was not very different from its Mesopotamian and Egyptian counterparts. Yet our image of a Greek mathematician is that of an elderly man occupying himself with the pursuit of knowledge for its own sake. Arguably the only element of truth in the previous sentence is the word *man*. The schools of the Ionian seaboard, reputed to be the first Greek scientific academies, developed their theories in close association with practice. The example that perhaps comes closest to reality is that of Archimedes. The image that is portrayed of him is, for that matter, comparable to that of a Renaissance engineer-cum-mathematician à la Galileo or Stevin.

Archimedes was what we might call an engineer, involved in, among other things, the design and construction of military equipment. In a sense he was also a mercenary, selling his formidable intellectual powers to the ruler of Syracuse. According to one account, he was killed as he tried to approach a Roman camp with some of his inventions after the fall of Syracuse. If true, then never in the field of human history has so much been lost to so many by one overzealous guard.

There are other examples of the close link between mathematics and engineering in ancient Greece: Eupalinos of Magara tunnelled through the Kastro hill on the isle of Samos, digging from both sides and meeting in the middle. The engineering tradition is also attested by Philo of Byzantium, who appears to have discussed military applications of catapults with the rulers of Alexandria. Heron and Diophantos, despite the seemingly theoretical appearance of the latter's work, fit seamlessly into this tradition.

Socrates: I heard, then, that at Naukratis, in Egypt, was one of the ancient gods of that
      country, the one whose sacred bird is called the ibis, and the name of the god
      himself was Theuth. He it was who [274d] invented numbers and arithmetic
      and geometry and astronomy, also draughts and dice, and, most important
      of all, letters.

[44] See E. ROBSON(2005).

[45] W. KNORR(2004), p.355, J. FRIBERG(2006).

[46] J. MANSFELD(1998), pp.83-84, 90.

1.3 The Greek written heritage

Despite our seemingly extensive knowledge of Greek mathematics, we have fewer original documents dating back to the Greek era than to Babylonian times. This is due to two factors: firstly, like the Egyptians, the Greeks wrote on degradable papyrus; and secondly, the large libraries where Greek knowledge was stored were destroyed in the course of history. The truth of the matter is, therefore, that today we possess just a few scraps of papyri testifying to ancient Greek mathematics.

The Greeks wrote on papyri in columns. They used non-accentuated capital letters without spacing between words. There were no fixed rules for hyphenating at the end of a line. Accents or other indications to clarify ambiguities were only slow to develop during the Alexandrian period. Accentuated Greek, with small and capital letters and with word spacing, was not introduced until the Byzantine period. It should therefore be kept in mind that, when placing accents, subsequent scribes may have had to choose between possible alternative readings, implying that we, too, are in fact reading an interpretation of the original.

Papyrus scrolls tore easily, yet they remained in use for a very long time. The earliest codices date back to the second century A.D., the most recent to the sixth century. A codex consisted of foliated papyrus leaves or (from the fourth century) parchment.

No complete papyrus scrolls from the Greek mainland are known, apart from some 'baked' copies[47]. The fact that we nonetheless know so many works from Antiquity is thanks to three factors. First and foremost, many scrolls were copied and preserved as codices in Byzantine and European convents and abbeys. Not only were the Greek versions of the Bible and philosophical treatises copied, but so too were numerous other works, covering a variety of topics. The only (almost) fully-preserved copy of an Ancient cookery book, for example, was found at the Abbey of Fulda[48]. Secondly, after the conquest of the North African coasts by the Arabs, many of the Greek treatises were translated into Arab. By about 1000, these Arab works had also been translated into Latin. Lastly, during the Humanist revival of the fifteenth century, which almost coincided with the advance of the Turks toward Constantinople, the Italians purchased numerous Greek codices in the shrivelling Byzantine Empire. The Turkish advance also prompted many Byzantine scholars (who may be regarded as the direct heirs to the Graeco-Roman tradition) to flee to Italy and elsewhere in Europe.

The earliest known Greek treatises date back to the fourth or fifth centuries. More often than not, we encounter them on *palimpsests*. Papyrus was a very precious material, so when a text had lost its relevance, it was sometimes erased in

---

[47] A 'baked' copy is a carbonized scroll that can be scrutinized scientifically, albeit with considerable difficulty. Moreover, much of the information contained in the documents is generally lost in the process. See M. GIGANTE(1979), G. CAVALLO(1983).

[48] See among others N. VAN DER AUWERA & A. MESKENS(2001).

order that the papyrus could be reused. Traces of the original texts remain visible, though, and can be made legible by modern techniques.

1.4   Numbers in Classical Greece

The ancient Greeks (or better: those who wrote Greek) used Herodianic and Ionic numerals side by side. The former notation is a mixed additive-multiplicative system, the basic symbols in which are:

$$| = 1, \Gamma = 5, \Delta = 10, H = 100, X = 1000, M = 10000$$

The symbol for one is simply a stroke, as in many scripts. The other symbols are the first letters of words. Γ is not a gamma, but an older representation of *pi*, as an abbreviation for *penta*, five. Δ stands for *deka*, H for *hekaton*, X for *kiloi* and M for *myrioi*.

Numbers were generally written additionally, as in $X\Delta\Delta\Delta = 1030$, but occasionally also in an multiplicative fashion, e.g. $\Gamma^{\Delta} = 5.10 = 50$.

In this way more complex numbers can be written, e.g. $XX\Gamma^{H}\Delta\Delta\Delta\Gamma|| = 2637$. The system was widely used throughout the Greek world and it is commonly encountered on amphorae[49].

The Ionian system is semi-positional. Letters of the alphabet are used to represent figures and multiples (< 10) of powers of ten.

1	2	3	4	5	6	7	8	9
α	β	γ	δ	ϵ	$\digamma$	ζ	η	θ

10	20	30	40	50	60	70	80	90
ι	κ	λ	μ	ν	ξ	o	π	$\koppa$

100	200	300	400	500	600	700	800	900
ρ	σ	τ	υ	φ	χ	ψ	ω	$\sampi$

1000	2000	3000	4000	5000	6000	7000	8000	9000
$,\alpha$	$,\beta$	$,\gamma$	$,\delta$	$,\epsilon$	$,\digamma$	$,\zeta$	$,\eta$	$,\theta$

As there are only twenty-four characters in the Greek alphabet, older characters were used for the remaining three figures: $\digamma$ = stigma/vau = 6, $\koppa$ = koppa = 90 and $\sampi$ = sampi = 900.

---

[49]See M. LANG(1956), next to M. N. TOD(1913) and (1979), about variants in the inscriptions on the various islands.

From the sixth century B.C. onward, the two systems were used side by side. Which of the two was most prevalent at one time would appear to have depended on whoever was the most dominant power: the Ionian cities or Athens. The earliest reliable evidence of the use of the Ionian system is on a *crater* dating back to around 575 B.C., the beginning of a period of Ionic cultural dominance that would last for about a century. Subsequently, between 475 and 325 B.C., the numerals were used only sporadically. The system did not disappear altogether, but it became marginalised when the Herodianic notation was readopted, as Athens gained in influence and Ionia's supremacy waned. Then, from the Alexandrian period onward, the Ionic numerals came in vogue again: they were preferred over the acrophonic numerals throughout the Greek world, with the exception of the Athenian polis. The Ionian system would remain in use in the Greek-speaking world until the fall of the Byzantine Empire. Stephen Chrisomalis asserts that the notation originated in Asia Minor as early as the sixth century B.C.[50] He further argues that the structure of the system was directly derived from the structure of the demotic numerals. These were in general use in Egypt in the sixth century B.C., when Ionian traders set up an *empórion* at Naukratis in the western Nile Delta (ca. 625 B.C. – see p. 2).

In principle, the system could cope with numbers up to 999. Multiples of a thousand were represented by the addition of a dash in front of the letters. Larger numbers were written using a mixed Ionic-Herodianic system,

e.g. $\overset{\beta}{\mathrm{M}} = 2.10000 = 20000$, or as β.

Here are some further examples:

$$\kappa\delta = 24, \upsilon\nu\beta = 452, \tau\eta = 305,$$
$$\overset{\kappa\beta}{\mathrm{M}}, \alpha\chi o\gamma = \kappa\beta., \alpha\chi o\gamma = 221673$$

Note that there is no place holder if a certain power of ten is not present.

Like the Egyptians, the Greeks wrote their fractions as a sum of unit fractions. Later, they began to write them as coupled numerals, whereby the distinction between nominator and denominator could be represented in different ways. For example, $\frac{2}{3}$ was written as $\beta'\gamma''\gamma''$ or $\frac{\gamma}{\beta}$.

The latter notation is not a fraction with nominator and denominator, but rather an abbreviation[51]. This remark is not without importance for the study of Diophantos, for it is not at all clear whether or to what extent the Diophantine corpus has been affected by such corruptions.

---

[50] S. CHRISOMALIS(2003)

[51] See D.H. FOWLER(1987), pp.263-268. Fowler suggests that these abbreviations provided the inspiration for the new notation of the Italian abacists and that they, through the use of later copies, corrupted our view of Greek arithmetical notation.

Although, traditionally, historians consider Greek mathematics to begin with
Thales, the earliest mathematicians where we find traces of algebraic reasoning are
Pythagoras and his followers. It is in their work that our notion of rational num-
bers originates.

The Pythagoreans attributed most of their ideas to Pythagoras himself, making
it virtually impossible to separate Pythagoras' teachings from later additions. It
would appear Pythagoras (ca. 570-500 B.C.) fled the island of Samos either because of the tyranny of Polykrates or for fear of the Persians. It was the beginning
of a peripatetic life that would lead him to Egypt, Crete and Delos (ca. 525 B.C.).
He eventually opened his school in Kroton, a Greek colony on the shores of southern Italy.

The Pythagoreans assumed that all natural phenomena could be expressed as
numbers. It was one of the greatest achievements of the Pythagoreans to recognize and emphasize that mathematical concepts, like number and geometrical
constructions, are in fact abstractions, ideas of the mind, which have to be sepa-
rated from the real-world objects[52]. It is to Plato that we owe the philosophical
notion of mathematical ideas existing independently from reality. When considering a triangle, we are in fact considering the creation of an Idea of the ideal or
essential triangle, which no drawn triangle can emulate. It is only in a Platonic
view that a triangle really exists. Any triangle we draw is disfigured by the fact
that each point and each line has a certain thickness, thereby shattering the very
concept of a triangle.

The Pythagoreans depicted numbers as pebbles or dots in the sand. It seemed
natural to them to associate a point or a pebble with the number 1. It was with
pebble arithmetic, or ψῆφοι, that mathematicians tried to solve summation problems[53].

They would then classify the numbers according to the shapes of the pebbles.
The numbers 1, 3, 6, 10, ... were referred to as triangular, because they can be arranged into a triangular shape. 1, 4, 16, 25, ... were, for obvious reasons, regarded
as square numbers. The Pythagoreans thus noticed that the sum $1 + 2 + 3 + \ldots + n$
leads to a triangular arrangement.

A standard method for obtaining summations seems to have been to make config-
urations with gnomons. For instance, it is obvious that, when equilateral gnomons
of 1, 3, 5, 7, ... are put together in a ψῆφοι-configuration, their sum will be (see
fig.1.2 left)

$$\sum_{i=1}^{n}(2i + 1) = (n + 1)^2.$$

---

[52]On Pythagorean philosophy see e.g. R. NETZ(2005)

[53]For detailed descriptions, see T.L. HEATH(1981), H.J. WASCHKIES(1989). In one cuneiform tablet (AO6484) from Seleucid times (dated early second century B.C.), similar algorithms
are used, indicating the practice may have been widespread. See J. HØYRUP(2000b), pp.3-6.
There is, however, no reason to suppose that the practice originated in Mesopotamia.

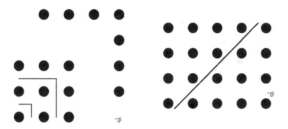

Figure 1.2 $\sum_{i=1}^{n}(2i+1) = (n+1)^2$ *(l) and* $\sum_{i=1}^{n} 2i = n(n-1)$ *(r)*.

On the other hand

$$\sum_{i=1}^{n} 2i = n(n-1).$$

Looking at the figure (fig. 1.2 right), which is composed of two equal triangular numbers (in this case 10) that are arranged into a rectangle with surface area $n(n+1)$ (here 4.5), it can be deduced that

$$\sum_{i=1}^{n} i = \frac{1}{2}n(n+1).$$

Similarly, they noted that square numbers are the sum of two consecutive triangular numbers, an observation that can be generalized:

$$\frac{1}{2}(n-1)n + \frac{1}{2}n(n+1) = n^2.$$

It is however doubtful that the Pythagoreans ever proved this generalization.

The Pythagoreans also looked for a rule to identify what we call Pythagorean triples. A Pythagorean triple consists of three numbers which are the sides of a right-angled triangle. This implies that they were at least aware of Pythagoras' theorem. They knew that for n odd $\left(n, \dfrac{n^2-1}{2}, \dfrac{n^2+1}{2}\right)$ is just such a triple[54].

This can be easily shown using ψῆφοι-arithmetic.
In the proof of $\sum(2i-1) = n^2$, all gnomons represent an uneven number and each gnomon (every odd number) is used exactly once when constructing a square.
Moreover, a square of an odd number is an odd number.

---

[54]The ascription of this method to the Pythagoreans is due to Proklos. See T.L. HEATH(1981)I, p.80

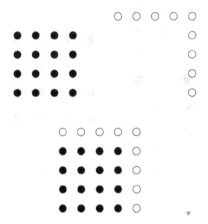

Figure 1.3 *Pebble-arithmetic proving $3^2 + 4^2 = 5^2$.*

Therefore, the square of an odd number n can be made into a gnomon. This gnomon can be added to a ψῆφοι-square with side $\dfrac{n^2 - 1}{2}$, making it a square with side $\dfrac{n^2 + 1}{2}$, which proves the property.

In this way some solutions for the equation $x^2 + y^2 = z^2$ are found, though not all, as there is in fact an infinitude of solutions. Nonetheless, it is more than ψῆφοι-arithmeticians may have hoped for: starting from a particular case, they found a general way for identifying solutions. These solutions are of course integer solutions to the equation.

Now what would happen in a special case[55] such as $2x^2 = z^2$? The equation says that there are two equal squares with side x that, when put together, make another square with side z. Obviously $x < z < 2x$ (since $x^2 < 2x^2 = z^2 < 4x^2$). Then $z = x + u$ for some positive u smaller than x and $2x^2 = (x + u)^2$. In ψῆφοι-arithmetic, this would mean laying out two squares (see fig. 1.4) . A square is subtracted from one of the given squares in such a manner that a gnomon of width u remains. This gnomon is laid next to the other initial square. Obviously this will not fit, as two squares with side u are lacking. The lacking part will then need to be constructed out of the pebbles of the subtracted square with side $x - u$. Therefore $2u^2 = (x - u)^2$.

We conclude that, if there is a solution (m, n) to the equation $2x^2 = z^2$, there will be an integer number, smaller than m, for which twice the square is again a square. Therefore, there are infinitely many numbers between m and 1 whose square is again a square. Obviously this is impossible. Consequently the equation cannot be solved.

---

[55] For a detailed discussion, see H.-J. WASCHKIES(1989), pp.272-275.

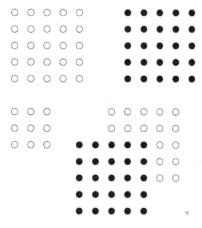

Figure 1.4 *Pebble-arithmetic proving the impossibility of $2x^2 = z^2$.*

It is easy to understand the procedure:

We know that
$$
\begin{aligned}
m^2 &= m^2 - 2mu + u^2 + 2mu - u^2 \\
&= (m - u)^2 + 2mu - u^2
\end{aligned}
$$
Now if $2m^2 = n^2$ then $n = m + u$, for some $u < m$
because $m^2 < 2m^2 = n^2 < 4m^2$.
Then:
$$
\begin{aligned}
2m^2 = m^2 + m^2 &= m^2 + (m - u)^2 + 2mu - u^2 \\
&= (m + u)^2 \\
\Rightarrow \quad m^2 + (m - u)^2 + 2mu - u^2 &= m^2 + 2mu + u^2 \\
\Rightarrow \quad (m - u)^2 &= 2u^2
\end{aligned}
$$
Which again proves the proposition.

This Pythagorean pebble arithmetic would develop into ἀριθμητική or the theory of numbers, which is to be distinguished from λογιστική or the art of calculation. In Plato, this distinction is not always clear. He frequently refers to both, as if they were overlapping parts of the same mathematical field. At other times, he draws a fine line between them. By the first century B.C., theoretical logistic had come to refer to the subfield of mathematics that occupies itself with artificial calculations about numbered collections[56]. Problems of logistic might have dealt with cattle weights, for example, and the most complex would involve indeterminate equations of the first degree[57].

---

[56] D.H. FOWLER(1992b), pp.105-117.
[57] T.L. HEATH(1981)I, pp. 12-16.

Another way of looking at numbers is to consider them as expressions of the length of –straight– lines (or any other quantity associated with geometrical figures, e.g. area, volume...). In this sense, it is natural that the Pythagoreans considered only positive integers as numbers. A fraction is a ratio of a length. Such ratios were called *commensurable* if both lengths could be expressed by a common unit. According to legend, philosophical problems presented themselves when the Pythagoreans discovered that some lengths are *incommensurable*[58]. Some scholars have argued that this issue was central to Greek mathematics, which had up to that point identified number with geometry. With the discovery of incommensurables, this identification was supposedly shattered. The Greeks therefore restricted the consideration of numerical ratios to commensurables, until a satisfactory theory of proportions was provided by Eudoxos. However, there are very few reasons other than the preoccupations of contemporary philosophers to believe there actually was ever a *Grundlagenkrise*.

According to Aristotle, the Pythagoreans used a *reductio ad absurdum* to prove the incommensurability of $\sqrt{2}$ and 1. It is open to debate whether this is true or merely an apocryphal attribution. Be that as it may, the proof is now a classic case for anyone studying the nature of numbers[59].

Let the ratio of the hypotenuse to the perpendiculars of an isosceles right triangle be $\dfrac{a}{b}$ and let a and b be the smallest possible numbers with which this ratio can be expressed. Therefore $\gcd(a,b) = 1$.
According to the Pythagorean theorem $a^2 = 2b^2$.
So a^2 is an even number, from which a is an even number, because the square of an odd number is odd.
Now $\gcd(a,b) = 1$ so b has to be odd. Let $a = 2c$ then $a^2 = 2b^2 = 4c^2$, so $b^2 = 2c^2$, which makes b even. But we proved that b is odd, therefore there is a contradiction.

Although Plato and Aristotle are regarded as the foremost Ancient philosophers, their direct influence on mathematics is modest. Still, their indirect impact is not to be underestimated. In a number of books, Plato stresses the importance

---

[58]The story that the Pythagoreans discovered incommensurability seems to originate with Iamblichus, *On the Pythagorean Life* 18(88) and a scholion to Euclid's tenth book: "It is well known that the man who first made public the theory of irrationals perished in a shipwreck in order that the inexpressible and unimaginable should ever remain veiled. And so the guilty man, who fortuitously touched on and revealed this aspect of living things, was taken to the place where he began and there is for ever beaten by the waves." (see T.L. HEATH(1981)I, p.154).

[59]Plato, *Thaetetus* 147d:

Theaetetus:Theodorus here was drawing some figures for us in illustration of roots, showing that squares containing three square feet and five square feet are not commensurable in length with the unit of the foot, and so, selecting each one in its turn up to the square containing seventeen square feet and at that he stopped. Now it occurred to us, since the number of roots appeared to be infinite, to try to collect them under one name, [147e] by which we could henceforth call all the roots.

Aristotle, *Prior Analytics* I.234^a 23-30, A. WASSERSTEIN(1958).

of the study of mathematics. Legend has it that the entrance to Plato's *Academy* bore the inscription: '*Let no one ignorant of geometry enter here!*'.

Eudoxos was undoubtedly the most important mathematician of Plato's Academy. To him we owe the theory of ratios of commensurable and incommensurable quantities. Eudoxos also introduced the concept of *magnitude*. It stood for notions such as angle, length, time ... , which can vary continuously but which are finite. What he in fact did was to avoid the concept of number, which, unlike magnitude, is a discrete set. His theory laid the foundations for manipulating incommensurable magnitudes and for breakthroughs in geometry. However, on the basis of the available evidence, we cannot but conclude that Eudoxos did not consider the concept of incommensurability[60]. Eudoxos did not assign a numerical value to a magnitude, but instead defined a ratio of magnitudes and a proportion. A proportion is an equality of two ratios, which thus allowed him to use commensurable and incommensurable magnitudes. Despite no numerical value being assigned to ratios, the approach made it possible to compare different ratios. Eudoxos' starting point was that two magnitudes have a ratio if a multiple of one of the magnitudes can be found that is larger than the other (i.e. $a : b$ exists if there are numbers m, n such that $ma > b$ and $a < nb$). This proposition is one of Euclid's definitions in book V and one of Archimedes' axioms.

The Eudoxan theory can be found in *Elements* V[61]. In this book, Euclid defines when four magnitudes are proportional or, as we would put it, when two ratios are equal. The beauty of this definition is that it is not necessary to consider the proportions of the same kinds of magnitude. Thus a and b may be the volume of a sphere, while c and d are the cubes of their radii.

> If a, b, c and d are four given magnitudes, and if m and n are positive integers, then the ratio $a : b$ is equal to the ratio $c : d$ if
>
> 1. If $ma > nb$ then $mc > nd$
>
> 2. If $ma = nb$ then $mc = nd$
>
> 3. If $ma < nb$ then $mc < nd$
> If there exist whole numbers m and n such that $ma > nb$ and $mc < nd$ then $a : b$ is larger than $c : d$.

Obviously this definition only makes sense in the case of incommensurables. If the magnitudes are commensurable, then condition (2) suffices, as both ratios would then be equal to the rational number n/m. The subtlety of the argument is that (2) is never satisfied for incommensurate magnitudes and that the equality then follows from (1) and (3).

---

[60]D.H. FOWLER(1994), p.224.

[61]According to a scholion to book V, "[s]ome say [the general theory of proportion] is the discovery of Eudoxos". T.L. HEATH(1956)II, p.112.

Consider, for instance, $\dfrac{\sqrt{2}}{1} = \dfrac{\sqrt{6}}{\sqrt{3}}$. It is clear that there are no integers that satisfy the relation $m\sqrt{2} = n.1$, which also means that $m\sqrt{6} = n\sqrt{3}$ cannot be true.

These Eudoxan definitions provide a solid basis for deducing further number theory theorems, often in a geometric fashion, without having to revert explicitly to irrational numbers.

Therefore it is doubtful that a genuine 'horror irrationalis' ever materialized. Diophantos, on the other hand, would shy away from incommensurables or irrational numbers. Although we shall make some comparisons to show the marked difference in style and content, we shall not go into the theory of numbers as propounded by, for example, Euclid, because this particular topic has no bearing on the issue at hand. Euclid's theory of numbers is geometric, not arithmetic, in nature. However, some of its terminology may be useful to us[62].

As we have already noted, magnitude (μέγεθος) is a finite continuous quantity. A number or integer (ἀριθμός) is a collection of units, but a unit is not a number. A ratio (λόγος) is a comparison of homogeneous quantities in respect of size. A proportion (ἀναλογία) is an equality of two ratios. Two magnitudes are commensurable (σύμμετρα) with each other if they have a common measure that divides each an exact integral number of times. Otherwise they are incommensurable (ἀσύμμετρα). In Euclid *Elements* X, lines are called expressible or rational if they are commensurate with a given line, whether in length *or* in square. That is to say, if the given line has length a, then both $2a$ and $\sqrt{2}a$ are expressible lines in terms of the given line. Also, the square of the given line is considered, and those areas that are commensurate with it are likewise called expressible[63]. In arithmetic treatises, such as the *Arithmetika*, the terms expressible and commensurable are used as synonyms.

1.5   All is number

It has been held that the Pythagoreans considered everything to be expressible as number. But what is number? This question shall present itself on several occasions in the course of this book.

Euclid and Diophantos tell us that number (*arithmos*) is composed of a multitude of units[64]. Such a definition is easy to understand if we refer to ψῆφοι-arithmetic,

---

[62]Based on W. KNORR(1975), p.15.

[63]The term *expressible line* was coined by David Fowler(1987), pp.167-172, to avoid having to translate with rational, where this term does not correspond with the present-day usage. The original meaning of ῥητός is "that which can be expressed". L. SÉCHAN & P. CHANTRAINE(1950), p. 1718. Also T.L. HEATH(1956) III, p. xx.

[64]Euclid, *Elements* VII, def 1 & 2: "Definition 1: A *unit* is that by virtue of which each of the things that exist is called one. Definition 2: A *number* is a multitude composed of units." Diophantos, I, introduction: "But also besides these things, as you know that all numbers are composed of some multitude of units, it is clear that (their) progression exists without bound...".

but what about rationals and irrationals? To be able to answer this question, we need to consider the meaning of the terms *logistic* (λογιστική, calculating) and *arithmetic* (αριθμητική, counting) in Greek *philosophical* thought[65].
Socrates, for example, says the following about arithmetic[66]

> [S]uppose someone asked me about one or other of the arts which I was mentioning just now: Socrates, what is the art of numeration? I should tell him, [451b] as you did me a moment ago, that it is one of those which have their effect through speech. And suppose he went on to ask: With what is its speech concerned? I should say: With the odd and even numbers, and the question of how many units there are in each.

And on logistic:

> And if he asked again: What art is it that you call calculation? I should say that this also is one of those which achieve their whole effect by speech. And if he proceeded to ask: With what is it concerned? I should say–[451c] in the manner of those who draft amendments in the Assembly–that in most respects calculation is in the same case as numeration, for both are concerned with the same thing, the odd and the even; but that they differ to this extent, that calculation considers the numerical values of odd and even numbers not merely in themselves but in relation to each other.

Despite the distinction made in this passage, Plato often treats the terms logistic and arithmetic as synonyms. In some cases he even introduces metric (μετρήτική), the art of measuring, in the same terms. How, then, should we interpret these notions when encountered in Greek texts? There is no consensus on this matter among philosophers, and perhaps not surprisingly so, as we are in fact considering texts spanning a period of almost a millennium, so that subtle changes in meaning are quite likely to present themselves. However, philosophers have tended to draw a line between theoretical and practical logistic. The former is seen to refer to that kind of artificial calculation that mathematicians tend to devise about number collections, such as cattle or ... pebbles. Obviously this would resemble number theory as we understand it. Practical logistic, on the other hand, is the arithmetic of merchants and craftsmen.
Theoretical logistic seems to be the origin of problems concerning the nature of numbers, primarily because the Greeks did not distinguish between the cardinal number of a set and the set itself. The members of the set were referred to as units, which added to the confusion with the number '1'.

---

[65]This paragraph is in large measure based on J. KLEIN(1992), pp.3-149, D. FOWLER(1987), pp.106-117 and M. CAVEING(1982), pp.758-796.
[66]Plato, *Gorgias*, 451a-c.

Now is it possible to divide the unit?

Theon of Smyrna notes that[67]:

> the one, when it is divided within the realm of sense, is on the one
> hand diminished as a body, and is, once the cutting up has taken place,
> divided into parts which are smaller than it, while on the other hand,
> as a number, is augmented.

The apparent paradox immediately holds the key to the division of unity. Whereas, after an object is divided, each part is smaller than the object itself, there are *several* pieces rather than one object.

The division of unity is therefore obtained, not by counting the object, but by counting the pieces. It would appear this is precisely what the Greeks did when considering the fraction $\frac{1}{n}$. They took a collection of $n + 1$ elements, because it is composed of unity, being the n-th part of n, and n. The '1' of the collection therefore is the n-th part of the remaining collection of n elements.

Obviously practical mathematicians were not concerned with these frivolities of philosophy, so they had little qualms about using fractions. After all, experience had told them that unity can indeed be divided: one *amphora* of wine makes many *glasses* of wine. . .

It has been argued by J. Høyrup[68] that mathematicians tend to care more about conquering new mathematical ground than about consolidating philosophically what they already possess. To put it bluntly, philosophy and practice are at odds. Diophantos, too, got caught in the middle. In his introduction, he asserts that *all numbers are composed of some multitude of units*, while in problem I.23 he refers to $\frac{50}{23}$ as *50 of 23rds* and $\frac{150}{23}$ as *150 of the said part*[69]. Moreover, in G IV.31, he merrily divides unity into two numbers, which obey certain relations. Mathematical necessity *always* wins over philosophy in Diophantos' *Arithmetika*.

---

[67] J.KLEIN(1992),pp. 39-40

[68] J. HØYRUP(2004), p.144.

[69] See p. 45 for the conventions on the numbering of the Diophantine books.

Chapter 2

Alexandria ad Aegyptum

2.1 The capital of memory

From the time of Alexander the Great, *Alexandria ad Aegyptum*[1], the Hellenic
capital of Egypt designed by Deinokrates, was a centre of learning.
It was situated at the westernmost tributary of the Nile, along its widest chan-
nels, and seemed destined after Alexander's destruction of Tyre to become the
dominant centre of trade between the Mediterranean, the Nile Valley, Arabia and
India. Ideally located at the crossroads of these cultures, it attracted people from
everywhere and could rightly be called the world's first cosmopolitan metropolis.
The city was oblong-shaped: about 4 miles at its longest and about a mile wide[2].
The streets were laid out in a checkerboard pattern, with two large thoroughfares
that almost bisected the city. Off the mainland, in the harbour of Alexandria, lay
the isle of Pharos with its famous lighthouse. The island was almost connected
to the shore by a long finger of land, a promontory known as Lochias, which
stretched out towards the east. A causeway called the Heptastadium was built
from the mainland, thus closing the bay toward the west.

The tongue of land on which the city stood made it singularly adapted to its
purpose as a commercial and military centre. Lake Mareotis, which bounded the
city to the south, served both as a wet dock and as a general harbour for inland
navigation along the Nile valley. Economically, it was the largest market in the
inhabited world[3] and the commercial hub of the Eastern Mediterranean. In the
pre-Roman era, the city's economy declined temporarily, but it would soon recover
and prosper again under the Romans. Egypt's agrarian economy was focused on
Alexandria, with its stockpiles of grain destined for export. Other export products

---

[1] Alexandria *by* Egypt, not *in* Egypt.
[2] Pliny, *The Natural History*, 5.11 and Strabo, Geography, XVII, I, 8.
[3] Strabo, *Geography*, XVII, I.13.

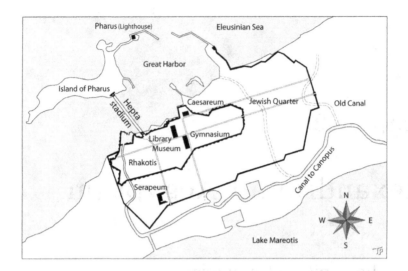

Figure 2.1 *Ancient Alexandria*

included papyrus, book scrolls, glasswork, jewelry, fragrance and medicine.
The harbour also offered military benefits: it was large enough to accommodate large fleets, while its narrow entrance made it easy to defend.

According to Strabo, Alexandria's famed salubrious atmosphere was attributable to its location, in between the sea and Lake Mareotis, which was filled annually by waters from the Nile[4]. Galen also believed that the regular replenishment of the lake prevented pestilences, which he associated with marshes of stagnant waters. However, the waters would eventually recede, creating a fertile breeding ground for disease-carrying mosquitoes. Obviously there is no reason to imagine that the autumnal outbreaks of *pestilential fevers* (= malaria) described by sixteenth-century local physicians, and which were associated with the propinquity of Lake Mareotis, might have been absent in Antiquity[5].

The population of Alexandria is assumed to have been close to half a million[6]. The city was divided into three districts: the Jewish quarter, the Greek quarter or *Brucheion*, and the Egyptian quarter or *Rhakotis*. After 31 B.C., the Romans, who occupied the Greek quarter, became the fourth ethnic group. The Jewish quarter had its own walls and gates, as hostilities between Greeks and Jews were common.

---

[4]Strabo, *Geography*, XVII, I.7 "When a large quantity of moisture is exhaled from swamps, a noxious vapour rises, and is the cause of pestilential disorders. But at Alexandreia, at the beginning of summer, the Nile, being full, fills the lake also, and leaves no marshy matter which is likely to occasion malignant exhalations".

[5]W. SCHEIDEL(2001), p.79.

[6]W. SCHEIDEL(2001), p.184.

The Royal or Greek quarter contained the most important public buildings, lush gardens and the Royal Palaces. It is here that the far-famed Library and Museum were located.

The Museum, or Temple of the Muses, is believed to have been erected under Ptolemy I (367-283 B.C.)[7]. It was a semi-religious research institution based on the model of the Athenian Academy and Lyceum and devoted to the cult of the Muses. Contrary to the Academy and the Lyceum, however, it was under government control. The Museum was apparently located within the palace walls[8]. Not all members of the Museum are known, nor whether the institution had a continuous existence. However, we do encounter some famous names among its librarians: Apollonios of Samothrace, Eratosthenes (the mathematician and scientist), Aristophanes, and Aristarchos of Rhodes[9]. The members were not obliged to teach, but it seems likely that most were surrounded by a group of students. It was at the Museum that the first attempts at literary criticism were made. The librarians were painfully aware that the texts they received were imperfect in various ways. Therefore, they indicated possible corruptions, standardized texts, and developed a number of reading aids, such as punctuation and accentuation. From 38 onward, membership was a kind of award bestowed upon civil servants and high-ranking officers[10]. There is no detailed account of daily life at the Museum at any time in its history. Indirect evidence, mostly relating to Kings and Emperors[11], comes from authors such as Strabo and Athenaeus, or from imperial biographies.

A covered marble colonnade connected the Museum with an adjacent stately building: the famous Library known as the *Alexandrina*[12].

---

[7]E.A. PARSONS(1952), pp.83 ff., P.M. FRASER(1972), pp.312 ff.

[8]"The Museum is also part of the Royal Quarters, having a public walk (*peripaton*), seating chamber (*exedra*), and a large building containing the dining-hall of the men of learning (*philologon andron*) who participate in the Museum. This group of men have common property as well as a priest in charge of the Museum, appointed in former times by the Kings, and nowadays by the Emperor." Strabo, *Geography*, XVII, I.8.

[9]E.A. PARSONS(1952), p.116, L. REYNOLDS & N. WILSON(1974), p.8, R. MacLEOD(2000), p.6.

[10]See for example N. LEWIS(1963).

[11]Emperor Claudius (10 B.C.-54 A.D.), for example, had a new wing added to the institution: "At last he [=Claudius] even wrote historical works in Greek, twenty books of *Etruscan History* and eight of *Carthaginian History*. Because of these works there was added to the old Museum at Alexandria a new one called after his name, and it was provided that in the one his *Etruscan History* should be read each year from beginning to end, and in the other his *Carthaginian*, by various readers in turn, in the manner of public recitations." Suetonius, *Life of Claudius*, 42.

[12]For an account of the history of the Library, see E.A. PARSONS(1952). Excellent brief overviews can be found in D. DELIA(1992), M. EL-ABBADI(1990/92) and K. STAIKOS & T. CULLEN(2004), pp.157-245. Other works, some of which tend to adopt a strong literary perspective, include L. CANFORA et al (1988).

The library of Alexandria was supposedly erected by Ptolemy II (309-247 B.C.), although no ancient account of its establishment exists[13]. Estimates of the size of the library collection vary from 400000 to 700000 volumes, though the distinction between books, scrolls and chapters is not always entirely clear[14]. The Library is said to have held Greek, Roman, Jewish, Persian, Ethiopian, Babylonian, Phoenician and Indian writings, and its collection seems to have grown so rapidly that part of it had to be housed in the temple of Serapis or Serapeum, in the Rhakotis district[15].

However, R. Bagnall [16] feels these numbers are grossly exaggerated. We know of only about 450 Greek authors, which, at an average of fifty scrolls each, makes 31250 scrolls. The cited figures are ten times greater, implying that we know just 10% of the classical authors. This, argues Bagnall, seems highly unlikely. On the other hand, the suggested physical space needed to store the number of scrolls is quite reasonable[17]. Also, we know that Seneca reproached Livy for showing regret

---

[13]Diogenes Laertius 4.1, 5.51. The Alexandrian library was certainly not the first of its kind. Assurbanipal, for example, had previously established a library in Niniveh, which at its height seems to have held some 30000 tablets. See also D.T. POTTS in R. MacLEOD(2000). Pergamum is said to have had a library once comparable in size to the Alexandrina, possibly containing up to 200000 scrolls. It is believed by some that competition with the library of Pergamum led to a veritable bookhunt. The Seleucids built a library in Antioch about which little is known. E.A. PARSONS(1952), pp.29 & 49.

[14]Flavius Josephus, *Antiquities of the Jews*, 12.11, Aulus Gellius, *Attic Nights*, 7.17.

[15]Tacitus, *History*, 4.84, Tertullian, *Apologeticum*, 18, Epiphanius, *De Mensuris et Ponderibus*, 11, J.S. McKENZIE, S. GIBSON & A.T. REYES(2004), pp.99-100.

[16]R. BAGNALL(2002)

[17]We can estimate the physical size of the Library from these numbers. Suppose that a papyrus scroll has a length of 10 m and a width of 30 cm. Further suppose that, when rolled up, the 'empty' part in the middle has a radius of 1 cm, and that the thickness of the papyrus is 1 mm. To simplify calculations, let us assume that the spiral formed by the papyrus can be approximated by a series of cylinders for which $R_{n+1} = R_n + 1$ [mm]. A scroll would then have a diameter of about 114 mm, or just over 10 cm. The surface area of the base of the cylinder formed by the rolled-up scroll is 0.0102 m². Assuming that about 70% of the storage space is actually taken in by the scrolls, then each scroll occupies an average of 0.0145 m² in space. Further assuming there are 400000 scrolls, we find the storage space would have to have an area of 5800 m². Imagine that the scrolls were stored on shelving measuring 10 m in length and 3 m in height, or 30 m². Dividing 5800 by 30, we arrive at 193 shelving units, which we round off to 200. Finally, assuming that a combination of two shelving units occupy a breadth of about 1 m (including sufficient space between them to allow access), we arrive at a floor area of 1000 m² or 10 m by 100 m. By comparison, at the presumed site of the Serapeum, 19 shelving rooms of about 4 m by 3 m were found. Under the same assumptions, we have 4×2 shelves, 3 metres deep. Multiplied by 19, we arrive at 1368. Divide this number by 0.0145 and we find nearly 95000 scrolls. Even if the figure was only about half this estimate, that is still a sizable collection. If the Serapeum was indeed a daughter library, then the number of volumes in the collection of the library as a whole may have run into six figures.

The figure of 50000 suggested here is very close to that mentioned by John Tzestzes, a twelfth-century Byzantine scholar, in a scholion to Plautus. He claims there were 42800 volumes in the outer library, and that the palace library contained 400000 mixed volumes [=unsorted] and 90000 volumes and digests. Tzestzes repeated these numbers in *Prolegomena to Aristophanes*. See E.A. PARSONS(1952), pp.106 ff.

at the destruction of 40000 (forty thousand) volumes in the Alexandrine wars[18], which suggests that the total collection must have been considerably greater. In 47 B.C., Caesar ordered the torching of the Egyptian fleet. Nearby warehouses and the (or a) library also caught fire and perished[19]. Mark Anthony allegedly had 200000 books transferred from the Pergamum library as compensation[20]. Estimates for the third century, when a large part of the library was possibly destructed, vary from between 200000 and 400000[21].

Whatever the size of the collection of the Library, its grandeur must have been awe-inspiring. Agents of Ptolemy III scoured the Mediterranean for books, an enterprise which would be repeated by Baghdad caliphs and Roman pontiffs alike. It made Alexandria pivotal in the ancient scholarly world, giving its scientist and literati an unparallelled access not only to Greek books, but also to Babylonian, Jewish and Egyptian writings. Among the scholars reputed to have visited or to have been invited to either the Museum or the Library are Euclid, Archimedes, Eudoxos, Aristarchos of Samos (the astronomer) and Hipparchos. Its reputation was still intact in the second century, when emperor Domitian sent scribes to Alexandria to copy books that had been lost in the Roman library[22].

Contrary to popular myth, the Alexandrian library was not destroyed by the Arabs. It was rather destroyed and rebuilt on several occasions. The most destructive event arguably took place in 272, when the civil tension that had always been present in the city escalated and turned violent. Alexandria's walls were torn down and the Greek quarter, with its Library and Museum, was left in ruins. The

---

[18]R.S BAGNALL(1993), p.351; "Forty thousand books were burned at Alexandria; let someone else praise this library as the most noble monument to the wealth of kings, as did Titus Livius, who says that it was the most distinguished achievement of the good taste and solicitude of kings." Seneca, *De tranquilitate animi*, 9.5.

[19]Amminianus informs us in his *Roman History* (written around 353-378) that 70000 books perished. Amminianus Marcellinus, *Historiae XXII*, 16.13. E. A. PARSONS(1952), pp.304-307, suggests that it was not the library that went up in flames but a warehouse, where 40000 scrolls had been stored for shipment to Rome.

[20]A. MEASSON(1994), pp.32-36. "Again, Calvisius, who was a companion of Caesar, brought forward against Antony the following charges also regarding his behaviour towards Cleopatra: he had bestowed upon her the libraries from Pergamum in which there were two hundred thousand volumes." Plutarch, *Antony*, 58.4.

[21]"So being asked in our presence how many myriads there are of books, he answered—'Over twenty myriads, O king: and I shall endeavour to have the rest made up to fifty myriads in a short time." Eusebius of Caesarea, *Praeparatio Evangelica*, 350b, "Ptolemaeus was a great lover of literature. With the help of Demetrius of Phalerum and other distinguished men, he used the royal funds to buy books from all over the world, and gathered them in two libraries in Alexandria. The outer library had 42,800 volumes; the library inside the palace complex had 400,000 mixed volumes, and 90,000 unmixed single volumes. Callimachus later compiled a catalogue of these books." Johannes Tzetzes, *Prologomena de Comoedia*, 20.

[22]"At the beginning of his rule he neglected liberal studies, although he provided for having the libraries, which were destroyed by fire, renewed at very great expense, seeking everywhere for copies of the lost works, and sending scribes to Alexandria to transcribe and correct them." Suetonius, *The Life of Domitian*, 20.

library's deathblow seems to have been dealt during the troubles of 391[23].

Unlike the rest of Egypt, which has a dry climate, Alexandria's climate is Mediterranean. As we have previously mentioned, papyrus does not preserve well in such conditions. Had the papyri not been replaced with copies on a regular basis, texts from the age of the Ptolemies would not have survived until the seventh century, or they would, at the very least, have deteriorated beyond repair or legibility. It is more likely that climate, bugs, mice and deterioration acted as a *slow fire* that gradually consumed the library holdings than that they were destroyed by Christian mobs or Arab conquerors.

Moreover, according to written sources, no fewer than twenty-three earthquakes struck the Egyptian coast between the years 320 and 1303, including a particularly severe one during the summer of 365. Over time, the harbour floor dropped more than 20 feet, so that the Royal Quarters, where the Museum was located, effectively collapsed and sunk beneath the waves.

Whether or not the Library was ever as grand as some ancient writers would have it, its fabled existence alone was enough to feed the imagination of notable book collectors during the Renaissance. It was this imagination that laid the foundation for the literary rebirth of many ancient writers, Diophantos included.

2.2 Diophantos' Alexandria

The only thing we can be reasonably sure of in the case of Diophantos is that he lived in Alexandria under Roman rule, most probably somewhere during the third century, though we cannot date him even to within several decades. Consequently, we do not know either whether he was ever required to pay the high taxes imposed by Augustus or whether he lived through or died during the pestilence that raged during the reign of Emperor Gallienus.

After Octavian had defeated Mark Anthony's forces at Actium and reconquered Egypt, he founded a new town in the Nile Delta, just east of Alexandria. It was called Nicopolis. Having bitter memories of Alexandria and Egypt, he imposed high taxes on their populations. He also put Egypt under direct imperial supervision, so that he controlled the food supply to Rome. His successors however would impose less harsh regimes. Under the Caesars, Alexandria was actually leniently governed, for it was in their interest to be popular in the city that commanded the largest granaries of Rome. The canal between the Nile and the Red Sea, which served a similar purpose as the present day Suez Canal, was redug. Goods from Asia were transported along the Nile to Alexandria, making it the world's principal commodities market. Most of the Caesars had some kind of relationship with Alexandria[24]. The first important change in their polity was introduced by the

---

[23]See also D.H. FOWLER(1987) and C. JACOB(1998).

[24]The city was favoured by Claudius, who added a wing to the Museum (see footnote 11). Claudius also had to deal with animosity between Greek and Jewish inhabitants of Alexandria.

Emperor Severus in 196. The Alexandrian Greeks were no longer formidable, and Severus accordingly restored their senate and municipal government. He also ornamented the city with a temple of Rhea and with a public bath, the *Thermae Septimianae*.

Alexandria did suffer terribly, though, after a visit from Caracalla in 215. Although he was greeted with hecatombs, he ordered the massacre of Alexandria's youth of military age in retribution for the fact that he had been mocked in the city in previous years [25].

Under Roman authority, Alexandria had previously enjoyed peace and stability. However, as imperial authority became more and more fragmented in the mid-third century, political stability in the city deteriorated. In the last quarter of the third century, Alexandria lost its predominance in Egypt. The native Egyptian population, reinforced by Arabian immigrants, had become a turbulent force. Diocletian's siege and subsequent capture of Alexandria in 298 seems to have been a watershed in the city's history. Throughout the autumn and Spring of 297/8, Diocletian, rather uncharacteristically, laid siege to the city in an attempt to crush the Egyptian rebellion centred around the cities of Alexandria and Coptos, who had backed the usurper Lucius Domitius Domitianus and his successor Aurelius Achilleus[26].

During the reign of Gallienus, Alexandria appears to have been struck by a pestilence[27], but it is not clear whether this was a particularly fierce outbreak of malaria or another infectious disease to which Egypt, and Alexandria in particular, was prone because of its position at the crossroads of civilizations.

Religiously, Alexandria was a curious mix, where Eastern and Western faiths met, crashed or blended[28]. Alongside Judaism, the cult of Serapis was widespread in Alexandria and indeed throughout Egypt. It was in itself an amalgam of religious practices, originating in the need to make Egyptian religious traditions more accessible to the Greeks. According to this cult, the sacred bull Apis, after its death, merged its divine characteristics with those of the god Osiris. In Alexan-

---

He warned them of the possible consequences if they forced the benevolent ruler to take action (H. BOTERMANN(1996), pp.107-114, B. LEVICK (1990), pp.89 and 182-185.). Nero intended to visit Alexandria, but never set sail, because of an ominous portent (Suetonius, *Nero*, 19). Alexandria also served as the headquarters of Vespasian (C. Tacitus, *Historiae*, 3.48) during the civil wars that preceded his accession. Struck by a dearth, the city was supplied with corn by Trajan (C. Plinius the Younger, *Panegyricus*, 31). And in 122, Alexandria was visited by Hadrian, who provided a graphic picture of the population (Vopiscus, *Saturninus*, 8).

[25]"Then he betook himself to Alexandria, and here he called the people together into the gymnasium and heaped abuse on them; he gave orders, moreover, that those who were physically qualified should be enrolled for military service. But those whom he enrolled he put to death,... ", Spartianus, *Caracalla* 6, see also Dion. Cassius LXVIII.22 and Herodianus IV.8-9.

[26]Eutropius IX.22

[27]Eusebius, *Historia Ecclesiastica* XXII, (ca. 263) W. SCHEIDEL(2001) does not record any particular pestilence for this period.

[28]On religion in Roman Egypt, see D. FRANKFURTER (1998).

dria, as we have seen, the temple of Serapis also served as a book repository for the main Library[29]. Although it remained an impressive structure, it endured alternating periods of prosperity and neglect. Around 181, the temple was destroyed by fire, in just one incident during which many manuscripts must have perished. It was later rebuilt on an even grander scale.

The presence of so many religions meant that unrest was never far away in this, arguably the most polyglot of Roman cities. Things did not improve when a new and unsettling religion made its appearance in Alexandria: Christianity. Christianity came to Alexandria relatively early, under the influence of, among other things, the presence of a large Hellenized Jewish community. It was supposedly introduced by St. Mark. From the time of Nero onwards, Christians would have to endure sporadic local and sometimes Empire-wide persecution. Alexandria, true to its reputation, saw the emergence of the first centres of Christian learning, such as the Cathechetical School, which –despite the persecutions– would continue to gain in influence in subsequent centuries[30]. The Christian Church began to thrive from around the mid-third century[31], but it was not until Constantine emerged as sole ruler that Christianity truly won the day.

Alexandria's position as the capital of the East was undermined when, in 324, Emperor Constantine decided to found a new city bearing his name. Constantinople would become the new seat of power in the Roman Empire. Alexandria's grain ships would no longer feed Rome, but the new capital.

So, assuming that Diophantos lived in the third century, he may well have been a witness to Caracalla's cruel treatment of Alexandria, the rise of Christianity, Diocletian's siege and the persecution of the Christians.

---

[29]On the cult of Serapis, other pagan religions and early Christianity see A.K. BOWMAN(1986), 167ff.

[30]P. SCHAFF, H. WACE & A.C. McGIFFERT (s.d.), p.345, footnote 1506.

[31]In 200, Severus' imperial edict forbade all subjects in the Empire to "make Jews or Christians" (i.e. to convert people to either Judaism or Christianity). After his death (211), the persecutions stopped and the Church grew in numbers and in wealth. Under Decius (249-251), the profession of Christianity was denounced as incompatible with the requirements of the state. The persecutions were put to an end after his death, only to resurface again under Valerian (257-261). After Valerian's capture by the Persians, his son Gallienus issued an Edict of Toleration (F. CONYBEARE(1914), see also Eusebius, *Historia Ecclesiastica*, III. 17 (Domitian), VI.1 (Severus), VI.28 (Maximinus), VII.1 (Decius and Gallus) VII.10 (Valerian), VII.13 &23 (Lucinius)).

Alexandria entered the late Roman period as the centre of a concerted rebellion against imperial authority, which had to be suppressed by the emperor Diocletian (284-306) himself. The beginning of the fourth century witnessed the start of Diocletian's 'Great Persecution' of Christians. It seems Egypt may have suffered more heavily than other areas, as one of the most fanatical anti-Christian persecutors, Sossianus Hierocles, held the office of prefect of Egypt (A.K. BOWMAN(1986), p. 45.). After Diocletian had retired from public life, a tetrarchy was organized, with ultimately Maximinus and Constantine as Augusti and Galienus and Lucinius as Caesars.

2.3   Education and the culture of *paideia*

Diophantos was undoubtedly part of, or at least very well acquainted with, the culture of *paideia*. Originally, the term meant "child-rearing", but it eventually became synonymous with "culture", the purely intellectual maturation and assimilation of the educational values acquired through schooling[32]. *Paideia* became a much more embracing concept, that could be understood as a code of behaviour, a way of life and of networking. It was acquired through an education that not only taught literature, but also allowed men of culture to master a behavioural code[33]. Regardless of their religious allegiance, men of standing were expected to participate in political and cultural life. Alexandria offered both, and at a high level of sophistication. The Museum and the Library naturally attracted prominent intellectuals, including scholars and authors, and many others used to send epistles or treatises to their friends and peers in the city. While these were not always intended for general circulation, it was common for the recipient to have them copied and sent to his circle of friends[34].
This culture did not change significantly after Christianity had come to prominence[35], but the new faith did add an ethos of hope.

Our general picture of education in Graeco-Roman Egypt is relatively clear[36]. However, the aspect of science and mathematics teaching is frequently ignored by authors on this topic.
For those not fortunate enough to receive private tuition, education began at an elementary school, which would not necessarily have been housed in a building. It may well have been in open air in the shadow of a tree. The goals of this education were modest, focusing on basic reading and writing skills. Arithmetic teaching followed the same basic structure it had done for over two thousand years. Pupils were familiarized with the basic operations: adding, subtracting, multiplying and dividing. Adding may have proceeded orally, by finger reckoning or by means of an abacus[37].
In specialized scribal schools, the acquisition of a deeper knowledge of multiplication and division was an integral part of the curriculum, as one might expect. Just as in Babylon, students also acquired metrological expertise: they were familiarized with weights, measures conversions and the monetary system. A book such as Heron's *Definitions* fits into this picture very neatly.

---

[32]R. CRIBIORE(2001), pp. 243-244.

[33]See E.J. WATTS(2006), p.2, pp.12-19.

[34]E.J. WATTS(2006), p.154.

[35]S. RAPPE(2001) on the incorporation of pagan elements into Christian education.

[36]This description is largely based on R. CRIBIORE(2001).

[37]Pupils were often required to recite simple additions in monotonous chants. G. CRIBIORE(2001), p.181. On finger reckoning, see B. WILLIAMS & R.S. WILLIAMS(1995); on the abacus, see R. NETZ(2002b).

Secondary education for the rich seems to have had a rhetorical focus. Students were taught the virtues and ideals to which one was expected to aspire. These were embodied in Greek and Roman literature alike. The Trojan hero Aeneas, for example, embodied the Roman ideals of duty and patriotism. The attempt of Greek Alexandria to emulate the Athenian model of education may have provided an incentive for the study of mathematics.

Higher education – at least during the Empire– seems to have been organized in the Musea. Although, as previously noted, members of the Alexandrian Museum were not obliged to teach, they did have students. Other Musea existed outside Alexandria, including at Ephesus and Smyrna. By the end of the fourth century, the word Museum had become synonymous with school[38].

Strikingly, most Greek mathematicians lived and worked in Alexandria: Euclid, Erasthotenes, Heron, Ptolemy, Pappos, Theon ... This may give rise to the rather misleading perception that there was a veritable concentration of mathematicians in the city. After all, we must not forget that we are looking at a timeframe of many centuries. Moreover, there is no evidence to suggest that there was any such thing as an Alexandrian school of mathematics[39]. It may just be one of those ironies of history that only texts by Alexandrian mathematicians did fortuitously survive[40].

While mathematicians enjoyed a high degree of social visibility, be it as land surveyors, artisans or indeed astrologers, they –and hence their texts– were equally clearly influenced by general social and cultural trends. One of these trends was that intellectuals and authors were becoming increasingly interested in classifications and rearrangements of previous knowledge, and they developed a predilection for commentaries and epitomes[41].

Despite the fact that many scholars believe that, by the fourth century, the teaching of mathematics had become either non-existent or limited to an elementary curriculum that was in every way subordinate to philosophy[42], it always remained part of the ideal programme of general culture. Indeed, the picture that emerges from the history of science is that, far from being invisible, mathematics was held in high esteem. The rise of Christianity did not change that. On the contrary, the new faith incorporated maths into its own educational programme. The theologian and teacher Origen (185-ca. 254), at the Cathechetical School, is known to have taught a Christian interpretation of physics, astronomy and geometry, whatever that may have encompassed.

---

[38] H.-I. MARROU(1948), pp.285-287.

[39] See for example G. ARGOUD(1994).

[40] See B.VITRAC(2008), p.531.

[41] S. CUOMO(2000), pp.48-56 and (2001), pp.249 ff.

[42] See for example M.L. CLARKE(1971) and D. PINGREE(1994). As already noted by S. CUOMO(2000), p.46, neither Euclid nor Nicomachus are elementary texts, yet they were part of the curriculum.

The teaching profession was not exclusively in male hands. Women in Hellenistic Egypt participated openly in society and tried to make a name for themselves in various professions. Some of the women known to have taught at the higher levels of education are Hermione, Agallis and Hypatia[43].

2.4 Heron of Alexandria: a Diophantine precursor?

Heron of Alexandria is one of the few known applied mathematicians of Antiquity. The name Heron was however rather common, so that it is hard to tell precisely which references are actually to Heron the mathematician. This makes him a rather elusive figure. Moreover, much of his work does not survive in its original form. It has been edited, altered and compiled so often that it is extremely difficult to distinguish Heron's hand from others, not to speak of the derivative or imitative works that are often attributed to him. This has resulted in an intricate web of more or less genuine and spurious manuscripts, and hence it should come as no surprise that the authorship of many of these works is disputed. To complicate matters further, some of his treatises have not been passed on to us in Greek. *Mechanica*, for example, survives only in an Arabic translation, while *Optica* is known to us only in Latin.

It is equally difficult to date Heron. Otto Neugebauer has argued that the 'recent eclipse', which Heron refers to, may be that observed in Alexandria on 13 March 62[44]. From this he concludes that Heron must have lived in the second half of the first century. This is corroborated by Dimitros Sakalis's research, which contains an in-depth study of words and phraseology used by Heron[45]. He also provides some indirect evidence, based on other sources. Galen seems to criticize Heron's work, without referring to him, but in the same phraseology[46]. Vitruvius (first century) does not mention Heron, although he refers to quite a few mathematicians and engineers. Moreover, he refers to mechanisms that are inferior to Heron's contraptions[47]. The oldest Hebrew geometry, the *Mishnat ha-Middot*, dating from the middle of the second century, was strongly influenced by the Heronian corpus[48]. Lastly, Proklos mentions that Heron was younger than Menelaos, who lived ca. 100.

In his works, Heron uses graecicized Latin words that only began to appear in the

---

[43]R. CRIBIORE(1996), pp.22-23.

[44]O. NEUGEBAUER(1938). However, N. SIDOLI (2005), pp. 250-252 puts it that Heron may have used a hypothetical eclipse. He argues that two other eclipses observed in Alexandria namely, in 133 B.C. and 3 B.C. tie in better with Heron's data.

[45]D. SAKALIS(1972).

[46]D. SAKALIS(1972), pp. 11-15.

[47]D. SAKALIS(1972), pp. 15-25.

[48]D. SAKALIS(1972), pp. 158-26.

first century[49], which reinforces the argument that he lived either in the second half of the first century or at the beginning of the second[50].

Heron's writings fall into many categories. He is, for example, one of our most important ancient sources on pneumatics, instruments and war engines. However, we shall restrict ourselves to his mathematical work[51], which, unlike his work on pneumatics, has received relatively little in-depth scholarly attention.

His writings reveal him to have been a well-educated mathematician, although his theoretical explanations are sometimes weak. But despite this shortcoming, he is an essential figure in the practical mathematics tradition that started in Babylonia. Furthermore, it would be a mistake to assume that practical mathematics was not an essential part of Greek mathematics as a whole.

Metrika is Heron's most important work. It is an introduction to practical geometry and measurement. Book I deals with plane geometry and builds freely on Euclid and Archimedes. The book is essentially constructed around three formulae: an iterative algorithm for the calculation of the square root of a number, the so-called Heronic formula[52] for the area of a triangle and the property that the area of a circle segment is larger than four thirds of the area of the inscribed triangle with the same base and height. Book II elaborates on the volumes of cones, cylinders, parallelepipeds, pyramids, frusta and Platonic solids. The volume of the sphere is determined to be two-thirds of the circumscribed cylinder. Book III discusses how figures can be divided into figures of a given ratio.

Definitions contains 133 definitions on geometrical terms, beginning with points and lines etc.

Geometrika would appear to be a different version of the first chapter of *Metrika*, founded entirely on exercises. Although it is clearly based on Heron's work, it is doubtful whether he was in fact the author.

Stereometrika deals with three-dimensional objects and is at least based on the second chapter of *Metrika*. Again, though, the original text is believed to have been altered considerably by later editors. Moreover, its two constituting books would appear to be different versions of the same work.

Mensurae is concerned with the measurement of a variety of objects. It is related to both *Metrika* and *Stereometrika*, but this book, too, is thought to be mainly

---

[49] For instance, πάσσον for *passus* and μίλιον for *milia*. See D. SAKALIS(1972), p.160.

[50] However, to illustrate the difficulty in dating Heron as a result of later additions, we also refer to the following examples. In *Geometrika* 21.26 and *Stereometrika* 1.21.3, a certain *Patricius* is mentioned. Patricius is identified as the Lydian expert in divination who was killed by Valens ca. 371 (T. MARTIN(1854), p. 220). In *Stereometrika* 2.54 we read: 'These [= the measures] were fixed under Modestus, who was praetorian prefect at the time.'. S. CORCORAN(1995) identifies Modestus as Domitius Modestus, who was praetorian prefect of the East under Valens from 369 to 377.

[51] The following description of the mathematical works is based on J.J. O'CONNOR & E.F. ROBERTSON (1999g).

[52] If A is the area of triangle with sides a, b and c and $s = \dfrac{a+b+c}{2}$ then $A^2 = s(s-a)(s-b)(s-c)$.

the work of a later editor.

The *Definitions*, like Diophantos' *Arithmetika*, are dedicated to a certain Dionysios, a very common name in Antiquity. Paul Tannery believed the two Dionysii to be one and the same person[53], but this was before Heron could be dated to the first century. Bearing this in mind, Markus Asper[54] concludes that Heron's Dionysios may be identified as Dionysios Glaukon. According to the Suida, this Dionysios was a student of Chaedemon, the Alexandrian librarian, whom he would succeed[55]. Dionysios would become a companion to all emperors from Nero to Trajan. He became the director of libraries and secretary responsible for correspondence, embassies and rescripts.

Of course, this identification will only stand if one accepts that *Definitions* was, at least in part, written by Heron. This attribution was already called into question by Hultsch and is further contested by Knorr[56]. On the grounds of style and genre, as well as the shared dedication to Dionysios, Knorr concludes that *Definitions* is closely associated with or may even have been written by ... Diophantos!

Indeed, most Heronian writings have prefaces that basically follow a consistent format. The preface to *Definitions* diverges from this pattern. However, like *Arithmetika*, it deals with the order of exposition and the pedagogical strategy[57].

Consider, for example, the closing thoughts of both texts (Knorr's translation):

Definitions: in this wise (houtôs) the subject matter will be well surveyable (eusynoptoi) for you.
Arithmetika: in this wise (houtôs) the elements will be well traversable (euodenta) for beginners.

If we accept Knorr's attribution and add to this Lucio Russo's thesis[58], the whole history of Euclid's *Elements* may be shattered. According to Russo, Euclid did not include the first seven definitions in his treatise, leaving fundamental entities undefined. In the Imperial Age, Euclid's choice could not be understood and the absence of definitions seemed to be a lacuna. As a remedy, Russo suggests, Heron first wrote his schoolbook *Definitions* and subsequently a list of Heron's work was compiled and inserted into Euclid's text. However, this would mean, at least in Knorr's view, that the Euclidean definitions are in fact ... Diophantine!

---

[53] P. TANNERY et al.(1912-1940)II, pp.535-538.
[54] M. ASPER(2001).
[55] Suida delta 1173, translated by Malcolm Heath.http://www.stoa.org/sol/
[56] W. KNORR(1993), esp. pp. 184-188.
[57] See also J. MANSFELD(1998), pp.55-57.
[58] L. RUSSO(2004), pp.320-327, esp. p.324.

It is in the Heronian corpus that we find some interesting indeterminate problems. These were not written by Heron, but added subsequently to *Metrika*. Heiberg included this collection of problems in his edition of *Geometrika*, creating the impression that they are genuine Heronian problems[59].

> *Find two rectangular areas such that the area and the perimeter are three times as large.*

The problem is equivalent to $\begin{cases} u + v = n(x+y) \\ xy = n.uv \end{cases}$

I do it like this, the cube of 3 is 27	n^3
which taken twice is 54	$2n^3$
If I deduct unity I find 53	$2n^3 - 1$
The first side thus is 53 feet	$x = 2n^3 - 1$
the other 54 feet	$y = 2n^3$
For the other rectangle 53 plus 54 is 107	
which multiplied by 3 [= 321,	$n(x+y)$
321 - 3] which is 318 feet	$n(x+y) - n = u$
One of the sides therefore is 318 feet	$u = 2n(2n^3 - 1)$
the other 3 feet	$v = n$
The area of the first is 954 feet	
of the other 2862 feet	

A possible explanation (as attributed to H. Zeuthen by T.L. Heath[60]) for this procedure is as follows: the problem is indeterminate, so start with a hypothesis, e.g. $v = n$.
Then $n(x+y) = n + u$, so u is a multiple of n, say nz and $n(x+y) = n + nz$ or $x + y = 1 + z$.
The second equation of the system yields:

$$\begin{aligned} xy &= n^3 z \\ xy &= n^3(x+y) - n^3 \\ xy - n^3 x - n^3 y &= -n^3 \\ xy - n^3 x - n^3 y + n^6 &= n^6 - n^3 \\ (x - n^3)(y - n^3) &= n^3(n^3 - 1) \end{aligned}$$

with an obvious solution $x - n^3 = n^3 - 1$ and $y - n^3 = n^3$,
which yields the solution:

$$\begin{cases} x = 2n^3 - 1 \\ y = 2n^3 \end{cases} \text{ and } \begin{cases} u = 2n(2n^3 - 1) \\ v = n \end{cases}$$

---

[59]Unfortunately, *Geometrika* was published in volume IV of Heron's works, while Heiberg explains his editorial method in volume V, thus adding to the confusion.
[60]T.L. HEATH (1981), pp.444-445.

Although algebraically correct, one may wonder whether this method was ever used by pre-Heronian mathematicians. It presupposes a large theoretical algebraic knowledge. Therefore, we suggest that it was found by means of another, most probably empirical, method.

Chapter 3

Diophantos and the *Arithmetika*

3.1 The manuscripts

Editing a classical text usually requires finding manuscripts and determining the variants and their interrelations. The aim of the editor is to approximate as closely as possible to the "original" through the comparison of existing manuscripts and papyrus fragments[1]. This is referred to as the *direct tradition*. In the case of Diophantos, this has been admirably done by André Allard in a hard-to-find edition[2]. Diophantos, as we intend to demonstrate, is an elusive figure about whom we now little more than that he probably lived in the third century. Just one complete work is definitely attributable to him. Another has been preserved only partially, and three further attributions are speculative. Clearly Diophantos, like Euclid, was a compiler, yet none of his sources are known. For some ancient writers, we possess more or less contemporary sources, that is to say papyri or ostraca from the Graeco-Roman period in Egypt, which can usually be dated to within about fifty years. One such group is closely associated with Euclid's *Elements*[3]. Another deals with commercial arithmetic or stereometry[4]. None, though, is in any way associated with Diophantos.

---

[1] S. ROMMEVAUX et al. (2001), p. 221. With reference to the *Elements*, Ken SAITO(2009) asserts (p. 810): "If we seriously want to argue about the content of the *Elements*, we should first try to establish the original text through such manuscripts as we possess; and if we cannot establish it, we should at least recognize the extent to which the surviving text resembles the original."
[2] A. ALLARD(1980).
[3] D.H. FOWLER(1987), pp.206-216.
[4] J. FRIBERG(2006), pp.193 ff.

A. Meskens, *Travelling Mathematics - The Fate of Diophantos' Arithmetic*, Science Networks.
Historical Studies 41, DOI 10.1007/978-3-0346-0643-1_3, © Springer Basel AG 2010

Moreover, we have to take into consideration the minusculization which every Greek text underwent in Byzantium[5]. The standard script for books in the Early Middle Ages was the uncial letter. Uncial script reached maturity in the fourth century and underwent little change after that. It was however slow and the size of the letters greatly restricted the quantity of text per page. This became problematic when relatively cheap papyrus grew quite scarce after the Arab conquest of Egypt and had to be replaced with much more expensive parchment. Not surprisingly, then, a new and more economical script was soon introduced. Known as the minuscule script, it had already been in use among scribes and accountants. Minuscule was compact and could be executed comparatively quickly. By the tenth century, uncial script was used only for special, liturgical books.

In the course of the ninth century, manuscripts in uncial script were transliterated into minuscule, but on a newly introduced material: paper. In this process of transliteration, numerous errors were introduced, primarily due the misreading of letters that were hard to distinguish in uncial script. This is not unimportant for texts such as Euclid's *Elements* or Diophantos' *Arithmetika* where a single letter, assigned to a point or number, can be crucial for a correct understanding. The largest part of the known Greek literature has come to us in transliterated form. In most cases, it is assumed that all extant manuscripts derive from a single archetypal minuscule copy from an uncial version, on grounds of, among other things, the observation that these manuscripts tend to contain the same errors.

It has long been recognized that the Greek manuscripts of the *Arithmetika* that have come down to us belong to two distinct classes: the Planudean and the non-Planudean class of manuscripts. The Planudean manuscripts contain scholia that were inserted by the thirteenth-century monk Maximos Planudes (see par. 4.3). Of the twenty-seven full manuscripts and four important excerpts that have been studied[6], none can withstand the test of mathematical rigour. Planudes even acknowledged this for his sources. Indeed, the manuscripts contain errors throughout, so that they arguably rank among the most corrupted Greek texts to have been passed down.

A stemma of the manuscripts[7], based on the work by Allard, is given in chap-

---

[5]On this process, see L. REYNOLDS & N. WILSON(1974), pp.51-54, on which this description is based.

[6]We refer to the last notable study by A. Allard, dating from the 1970s. Unfortunately, this work has never been published, apart from a very limited edition consisting in photocopies of his doctoral thesis. Its publication was announced by *Les Belles Lettres* but never actually materialized.

[7]Constructing a *stemma* is part of stemmatics or stemmatology, a rigorous approach to textual criticism. The stemma or "family tree", shows the relationships of the surviving manuscripts. The method works from the principle that "community of error implies community of origin". That is, if two witnesses have a number of errors in common, it may be presumed that they were derived from a common intermediate source, called a hyparchetype. Relations between the lost intermediates are determined by the same process, placing all extant manuscripts in a family tree or *stemma codicum* descended from a single archetype. This process of constructing a stemma is called recension.

ter 10. We shall refer to these various versions in due course, when discussing the era when they were produced or used.

Editors may also rely on other sources than direct descendants of the archetypal text, such as quotations by other authors, ancient commentaries or translations (mostly Arab). This is referred to as the *indirect tradition*. Adherents of this approach point out that Arab translations can be more faithful to the original Greek than subsequent Greek copies, because they are often based on older versions.
In view of the enormous amount of non-described and non-inventoried Arabic manuscripts, it should not come as a surprise that every once in a while an hitherto unknown manuscript is identified. That was the case in 1971, for example, when an Arab version of Diophantos was found. The discovery caused quite a stir, particularly when it emerged that the text contained content from hitherto unknown books. In these books, it becomes clear that Diophantos used the methods that he describes in his other works to their limits in order to solve higher-degree problems.
The study and analysis of the newly discovered Arab manuscript is the work of two Arabists: Roshdi Rashed and Jacques Sesiano[8].

Up to the 1970s, the indirect tradition in the case of Diophantos consisted in little more than a couple of quotes. The discovery of an Arab version of Diophantos made little difference in this respect however, as the Arab texts were entirely unknown in Greek. So, unfortunately, the Arab version can only be compared with apparent excerpts from the *Arithmetika* cited by other Arab authors.

Although Diophantos asserts that his treatise is divided into thirteen books, only ten of these books are known to us. The Greek manuscripts contain six books each, four further books are known to us in an Arab translation. The original order of these ten chapters would appear to have been G I, G II, G III, A IV, A V, A VI, A VII, G IV, G V, G VI[9]. The exact order of the missing three books within the series of thirteen is uncertain. In our description of the contents of the *Arithmetika*, we shall maintain the aforementioned order.

Having completed the stemma, the critic proceeds to the next step, called selection, where the text of the archetype is determined by examining variants from the closest hyparchetypes to the archetype and selecting the best ones. Where the editor concludes that the text is corrupt, it is corrected by a process called "emendation".
André Allard has published several articles on the descent of the Diophantos manuscripts (1979), (1981a and b), (1982-83), (1984).

[8] R. RASHED(1974), (1975) and (1984); J. SESIANO(1982). It was the keeper of the Iranian parliamentary library who initially drew Roshdi Rashed's attention to the existence of the text, which was kept at the library of Meshed. Rashed soon identified it as having been written by Diophantos. On the discovery of the manuscript, see R. RASHED(1984), LIX-LXXII; on the ensuing controversy, see A. ALLARD(1984) and (1987). The document is now kept in the library of Astān Quds, which is part of the mosque of the shrine of imam Rezā.

[9] J. SESIANO(1982), p. 5, R. RASHED(1984), p.VI. We shall refer to the first three Greek books simply as I, II, and III, because no confusion is possible.

3.2 Diophantos

As we have previously said, Diophantos (διοφάντου αλεξανδρέως) was either born in Alexandria or it was his permanent domicile. Although αλεξανδρέως was an appropriate term for a citizen of Alexandria, it was not necessarily synonymous. Therefore, we do not know whether Diophantos actually held Alexandrian citizenship[10].

Only two books can be attributed to Diophantos with any degree of certainty[11]: the *Arithmetika*, or *Arithmetic*, and a treatise on polygonal numbers. Three other titles are conjectured to have existed: *Porismata*[12], *Moriastika*[13] and *Arithmetika Stoicheisis*.

Unlike in most ancient mathematical texts, no reference is made in the *Arithmetika* to other mathematicians, which immediately rules out a posteriori dating of the treatise. An a priori dating is also very difficult, since the only author to refer to Diophantos is Theon of Alexandria[14]. Theon lived in the fourth century (ca. 335-ca. 405)[15], so we can safely date the *Arithmetika* to before 400. Whereas the *Arithmetika* was definitely written by Diophantos, the authorship of the book on polygonal numbers remains debatable (see par. 3.3). However, if we accept the attribution to Diophantos, then we are able to use the reference to Hypsikles, who is known to have lived around 150 B.C.

But even then it is impossible to date the life of Diophantos to within an interval of well over five hundred years, which more or less corresponds with the entire period of Roman rule over Egypt.

As R. Netz has previously noted, when mathematicians in Classical Greece cite peers, there is usually an age difference of no more than one generation[16]. With this knowledge in hand, and considering Theon's reference, we can tentatively put Diophantos' life around 300.

One of the earliest references to Diophantos, albeit with little biographical significance, is in the *Suida Lexicon*. The *Suida* is a tenth-century alphabetically organized encyclopedia containing some 30000 articles. The texts are based on earlier sources, but they are not always entirely trustworthy. Somehow, however, Diophantos' reputation must have lived on for hundreds of years, to the extent that his name became a byword for a logistic teacher.

---

[10]On Alexandrian citizenship see D. DELIA (1991). As in many other Greek cities, citizenship in Alexandria was hereditary. The only other means of obtaining citizenship was through naturalization. Citizenship would have brought many advantages, including exemptions from certain taxes.

[11]On Diophantos see T. HEATH(1964), P. VER EECKE(1926), J.D. SWIFT(1956) and N. SCHAPPACHER(2001).

[12]Mentioned in problems G V. 3, G V.5, G V.16.

[13]Mentioned in a scholion to Nicomachus' *Arithmetica*.

[14]T. HEATH(1964), p.2, P. TANNERY(1895) II, pp.35-36.

[15]J.J. O'CONNOR & E.F. ROBERTSON (1999e).

[16]R. NETZ(2002a), pp.215-216, see also B. VITRAC(2008), p.530 ff.

Under the lemma *A drachma raining hail* (δ 1491), we find[17]:

> In the case of Diophantos a drachma became the subject of spec-
> ulation. When the hail stopped at that minute (falling) from the upper
> air, they joked it was a handful (= drachma) of hailstones.

The word drachma has three meanings: a handful, an old Athenian coin and a
weight[18]. Referring to the sometimes unrealistic nature of problems of logistic, the
answer to the question "How much hail has fallen?" would appear to be a play
on those different meanings: "Why, a *drachma*.", i.e. a handful, which weighs a
drachma and is worth a drachma ...

One of the earliest references to Diophantos as a person is by the Byzantine
intellectual Michael Psellos (1018-1081(?)). Psellos wrote a large number of treatises on very diverse topics, including in the fields of philosophy, theology and the
sciences. In a letter[19], Psellos refers to the treatise *The Egyptian method for numbers*, written by Anatolios and dedicated to Diophantos. Tannery identifies this
Anatolios with Anatolius of Alexandria, the Bishop of Laodicea (on the Syrian
coast) around 270-280. This bishop is indeed known to have written mathematical
treatises, fragments of which have been preserved. He was a student of Dionysius
of Alexandria (the later Saint Dyonisius)[20]. If one concurs with Tannery that a
treatise can only be dedicated to a person if this person is still alive, then Diophantos must have lived in the third century, which for that matter ties in nicely
with our foregoing remark regarding references to him. Moreover, this Dionysius
may well be Diophantos' dedicatee.

However, this dating has been called into question by W. Knorr[21]. In *Definitions on geometrical terms*, an introductory comment on Euclid that is ascribed
to Heron, we encounter a reference to a book of the same name, i.e. *Arithmetika
Stoicheisis*. These definitions are likewise dedicated to Dionysius, which was not
an uncommon name in Antiquity[22]. J. Klein therefore proposes another interpretation of the Psellos fragment. He argues that Psellos is referring to the differences
in symbolism between Anatolios and Diophantos. In this way, the Anatolios reference becomes an a posteriori dating: Diophantos lived before him. He also argues

---

[17]δραγμὴχαλαζῶσα· ἐπιδιοφάντου τοῦεωρητιχὸν ἐγένετο δραγμή. ἐπεὶ δὲ ἐπέσχε χάλαζα τότε
ἀπὸ τοῦ ἀέρος, δραγμὴ ν αὐτὴ ν χαλαζῶν ἐπέσκωπτον. Translation by Robert Dyer, Adler number
delta, 1491 on http://www.stoa.org/sol/. The translator thinks the translation must be 'a handful
of hailstones'. His explanation of the expression seems to be a little mathophobic: whatever
Diophantos (the teacher) was teaching disappears as fast as a handful of hail melts away.

[18]J.L. HEIBERG (1912), Heron, *Geometrika*, p. 411.

[19]P. TANNERY(1895)II, pp. 37-42.

[20]W. KNORR(1993), p. 183, Eusebius, *Historia Ecclesiastica*, XXXII. According to Eusebius,
Anatolios was a man of great distinction and erudition. He wrote a work on the calculation of
the Easter date and the *Institutes of Arithmetic*, in ten volumes. Part of this work was inserted
in Heron's *Definitions* (138), presumably by a Byzantine scribe.

[21]W. KNORR(1993).

[22]Paula-Wissowa V has 166 lemmas under the name Dionysus.

that Anatolios' treatise is dedicated to *another* Diophantos[23].

On the basis of stylistic characteristics, Knorr concludes that *Definitions*, which is generally ascribed to Heron, may in fact be attributed to Diophantos. The texts are, moreover, both dedicated to a Dionysius.

According to Tannery, the two were one and the same person[24]. On this basis, he concludes that Heron and Diophantos lived around the same time. A similar suggestion is made by Heath[25], following Heiberg, although he identifies the Heronian Dyonisius as L. Aelius Helvius Dyonisius, the Roman prefect (*praefecturs urbi*) of 301. However, since these authors presented their arguments, it has been established by O. Neugebauer that Heron must have lived around 62, which rules out the identifications of these Dyonisii[26], unless Diophantos lived in the first century, which, at least according to W. Knorr, is not implausible[27]. On the other hand, if Dionysius *can* be identified with Dionysius of Alexandria then, Knorr concludes, Diophantos must have lived a generation earlier, around 240.

A counterargument against these propositions is that Diophantos is cited by neither Nicomachus (ca. 100), nor Theon of Smyrna (ca. 130) nor Iamblichus (late third century). The use of the word *leipsis* (see below) also suggest that Diophantos lived later. Its first recorded use is in the second century. The term *hyparxis*, which Diophantos uses in a mathematical context, appears quite frequently in philosophical treatises belonging to the Neoplatonic school of Alexandria (ca. 200).

Since 1500, more than a thousand years after his death, various authors have speculated about the life of Diophantos, identifying him as an Arab[28], a Jew, a converted Greek or Hellenized Babylonian. None of these characterizations stands up to critical scrutiny though[29]. Whether we like it or not, the reality is that we

---

[23]The keyword in the Psellos fragment is *heterôs*, which, according to Knorr(1993), p. 184, should be read as *heteroi*.

ὁ δὲ λογιώτατος Ἀνατόλιος τασυνεκτικώτατα μέρη τῆς κατ' ἐκεῖνον ἐπιστήμης ἀπολεξάμενος ἑτέρως [?] Διοφάντῳ συνοπτικώτατα προσεφώνησε.

The reading in Tannery (1895) II, pp.38-39, translates as: *Diophantos treated this very precisely, but the very learned Anatolios collected the most essential parts of this theory, as described by Diophantos, in another way.*

For his part, Klein (1992), pp. 244-246, translates as:... *but the very learned Anatolios collected the most essential parts of this theory in another way than Diophantos.* He devotes a whole paragraph to the possible translations.

Knorr (1993), on the other hand, thinks the fragment should be read as: *but the very learned Anatolius, who had collected the most essential parts of this theory, dedicated it to that other Diophantos.*

[24]P. TANNERY(1896-1912), pp. 535-538.

[25]T.L. HEATH(1964), pp.306-307.

[26]O. NEUGEBAUER(1938).

[27]W. KNORR(1993), pp.156-157.

[28]There may be some confusion here with Diophantus the Arab, Libanius' teacher, who lived during the reign of Julian the Apostate. See also Suida λ486 and S.N.C. LIEU in: R. McLEOD(2000), pp.129-130.

[29]Resp. by O. SPENGLER(1923), pp. 96-99, 770, P. TANNERY(1912-1940)II, pp.527-539, D.

know next to nothing about Diophantos, one of the most original mathematicians of Antiquity.

3.3 The book *On Polygonal Numbers* and the lost books

The book *On Polygonal Numbers* is often ascribed to Diophantos, but this attribution is by no means certain. Evidence is simply lacking.

The book could be described as a distant echo of the ψῆφοι-arithmetic. In the introduction, the author defines the notion of polygonal numbers and announces that he intends to demonstrate, among other things, how to construct a polygonal number of a given type, given its side. He goes on to explain that he intends to begin with some preliminaries. Unfortunately, only four theorems of the book survive. All four belong to the preliminaries and deal with arithmetical progressions. They are proved rigorously, using metrical geometry of the line.

The theorems are (in anachronistic terms):

If $(a_i)_{\{i\in\mathbb{N}\}}$ is an arithmetical progression with difference v then

$$8a_{j+1}a_j + a_{j-1}^2 = (a_{j+1} + 2a_j)^2$$

If three numbers exceed each other in the same way, then the octuple of the product of the largest and the middle, increased with the square of the smallest, is a square whose side is equal to the sum of the largest number and the double of the middle number

$$a_n - a_1 = (n-1)v$$

If a random number of numbers have the same difference, then [the difference] of the largest and the smallest is a multiple of that difference, which emanates from the number that is a unit smaller than the number of chosen numbers.

$$\sum_1^n a_i = \frac{n}{2}(a_1 + a_n)$$

If a random number of numbers have the same difference, then the sum of the smallest and the largest, multiplied by the number of numbers, is the double of the sum of the given numbers.

BURTON(1991/95), p. 223. See N. SCHAPPACHER(2001) for a discussion on the claims of these authors.

If $\sum_1^n a_i = s_n$
then
$$8s_n v + (v-2)^2$$
$$= ((2n-1)v+2)^2$$

If, commencing at unity, a random number of numbers have a same difference, then the sum of all numbers, multiplied by the octuple of their difference, and increased with the square of the number that is two less than the difference, is a square whose side, from which two units are deducted, is a multiple of the difference of the numbers, which emanates from a number which, augmented with unity, is the double of the number of all given numbers, including unity.

In the proof of the fourth proposition, it is stated that "in this fashion we have proved what Hypsikles says in a definition". If we accept the questionable attribution of this book to Diophantos, then this presents an opportunity for a posteriori dating of his life. After all, Hypsikles was a mathematician from the mid-second century B.C. In view of the reference, Diophantos could, at the earliest, have been his contemporary. The attribution to Diophantos is based on the fact that the treatise *On Polygonal Numbers* appears in all known *Arithmetika* manuscripts, albeit very partially in some cases[30].

In the *Arithmetika*, Diophantos refers on a number of occasions to his *Porismata* (G.V 3, 5, 16), which is believed to be a collection of propositions dealing with the properties of certain numbers. Hence, it would appear to have been a theoretical work underpinning the *Arithmetika*. The testimony of Al-Karajī makes this hypothesis plausible. He attributes the algebraic proof for the solution of a second-degree equation to Diophantos[31], yet no such proof can be found in the *Arithmetika*.

According to Proklos, however, "a porism is a theorem resulting directly from the proof of another problem or theorem". The question then becomes: from which theorems exactly are they deduced? J. Christianides[32] has therefore suggested that there may be yet another lost work, namely Ἀριϑμητικὴ Στοιχείσις, or *Elements of Arithmetic*. Moreover, a reference to this title is made in a scholion in Iamblichus' comments on Nicomachus' *Introductio Arithmetica*. This scholion further refers to the last problem of the first book, in which the harmonic mean is mentioned. In problem I.39 (the last known problem of book I) of the *Arithmetic*, the harmonic mean is not dealt with explicitly, which seems to imply that the title refers to a

---

[30] A. ALLARD(1982-83), pp.59-72.
[31] A. ANBOUBA(1978), p.71.
[32] J. CHRISTIANIDES(1991).

–lost– book by Diophantos[33]. W.C. Waterhouse[34] gives an interpretation of problem I.39 in which the harmonic mean appears. According to him, it had implicitly always been there. R.Rashed, on the other hand, argues that the reference is to the *Arithmetika* itself. Indeed, in an Arabic treatise, *al-Fakhrī* by Al-Karajī, dealing with problems extracted from Diophantos (see par. 4.2, p. 112), we encounter a sentence that is lacking from the Greek version and that refers to the arithmetical mean, which, according to Rashed, may have been accompanied by a sentence on harmonic means.[35].

A final work to have been atributed to Diophantos is *Moriastika*, mentioned in just one scholion to Iamblichus' commentaries on Nicomachus' *Introductio Arithmetica*[36].

Bernardino Baldi, in *Cronica de matematici*(1707), also mentions a treatise on *Harmony*. He asserts that this work is unedited, giving rise to the assumption that a manuscript copy may have been available in circles close to him. It is however likely that a treatise by another author was appended to Diophantos' work, occasioning a misinterpretation of authorship. Baldi further writes that Diophantos is the author of an *Astronomical canon*, which was commented upon by Hypatia[37].
The latter assertion is possibly based on an interpretation of the description of Hypatia's work in the *Suida* (see section 4.1), where we read: "She wrote a commentary on Diophantos, [and one of] the Astronomical Canon, and a commentary on the Conics of Apollonios"[38]. The interpolation by Paul Tannery has become generally accepted. If it is omitted, we indeed come close to Baldi's claim.

---

[33] J. CHRISTIANIDES(1991) and W.KNORR(1993).

[34] W.C. WATERHOUSE(1993).

[35] R. RASHED(1994). "If three numbers have an equal difference, then [the sum of] the outer numbers is equal to the double of the middle one." (p. 44)

[36] P. TANNERY(1895) II, p.72.

[37] F. WIERING(2000). One of the volumes in which Diophantos' *Arithmetika* can be found, Matritenis Bib. Nat 4678 (= N48), also contains an anonymous astronomical text and two astronomical tables (see A. ALLARD(1982-83), p.66). This or a similar collation may have led Baldi to assume that the astronomical text was also by Diophantos.

[38] M.A.B. DEAKIN(1994), pp. 237-238 and translation by C. ROTH, Adler number upsilon 166 on http://www.stoa.org/sol/ .

3.4   Symbolism in the *Arithmetika*

As is the case with most Greek treatises, we only know the *Arithmetika* through Byzantine and mediæval copies, implying that the original text has gone through a process of minisculization. We can therefore only guess as to which, if any, Diophantine symbols appeared as capital letters in the original documents[39].

Therefore, all comments on style and notation are first and foremost comments on the Byzantine versions. This filtering process, combined with the uniqueness of the *Arithmetika*, makes it virtually impossible to look further back in time than the Byzantine era. Only if we compare Diophantos with original Graeco-Roman papyri is it possible to reach more robust conclusions.

The most important characteristic of the *Arithmetika* is the system it uses for representing what we refer to as polynomials, with abbreviations for the unknown[40].

$\overset{\circ}{\mu}$	ς	δ^v	κ^v	$\delta^v\delta$	$\delta^v\kappa$	$\kappa^v\kappa$
unity	number	square	cube	square square	square cube	cube cube
(x^0)	(x^1)	(x^2)	(x^3)	(x^4)	(x^5)	(x^6)

What we know as polynomials would have been written in the following fashion:

1. The coefficients were represented in Ionian style, after the unknown or its power

2. All terms to be subtracted were written after ⟱

3. The terms to be added were written without a summation sign, before the ⟱

E.g.

$\delta^v\overline{\delta}\,\overset{\circ}{\mu}\,\overline{\kappa\epsilon}$ we would write as $4x^2 + 25$

$\delta^v\delta\overline{\alpha}\,\overset{\circ}{\mu}\,\overline{\omega}$⟱$\delta^v\overline{\phi}$ we would write as $x^4 - 50x^2 + 800$

$\kappa^v\overline{\beta}\varsigma\overline{\eta}$⟱$\delta^v\overline{\epsilon}\,\overset{\circ}{\mu}\,\overline{\alpha}$ we would write as $2x^3 - 5x^2 + 8x - 1$

---

[39]See for example the remark in this sense by R. NETZ(1999b), p. 43. On minusculization, see par. 3.1.

[40]In the Tannery transcription, capital letters are used. We prefer to follow Allard's transcription, which uses small letters. The translation of this notation has always posed problems. Heath (1964) and Tannery (1895) do not translate the text, but rather paraphrase it in modern mathematical terms, including modern mathematical notation. Ver Eecke (1926) resolves the abbreviations and writes *l'arithme* for ς. Meskens & Van der Auwera (2006) translate the text, but use an x for the arithmos and modern exponentiation for higher powers. Allard (1980), in his rare translation, is perhaps most faithful to the text, using Renaissance-like abbreviations such as N (number), Q (quadrat) and C (cube).

Signs similar to ⋔ seem to have been in general use for subtractions. In *Papyrus graecus Vindobonensis 19996,* we encounter the sigil ⌒ [41]. A similar sign would appear to have been used by Heron in his *Metrika*[42].

Diophantos used $\overset{\circ}{\mu}$ to refer to numbers. It is often regarded to be nothing more than a symbol to indicate the independent term[43]. Had $\overset{\circ}{\mu}$ been omitted, confusion may have arisen, especially with sloppy handwriting. Is ςιγ (overline omitted) equal to 23x or 2x + 3? Writing ς$\overset{\circ}{\mu}$ιγ resolves this problem. However, this interpretation is not shared by J. Klein[44]. He sees $\overset{\circ}{\mu}$ as an abbreviation of *monas* (unity), necessary in the notation as a consequence of the meaning of the word *arithmos*, i.e. a certain number of something. Klein refers to a similar, non-abbreviated word used by Heron. The *arithmoi* calculated by Diophantos are, according to Klein, numbers of pure units: "all numbers are composed of a certain number of units"[45]. In this interpretation, Diophantos may have viewed a monas as divisible in parts.

Diophantos' *arithmos* can therefore be regarded as a positive rational number. Irrational numbers –although they are present– and negative numbers are not considered to be *arithmos*.

On the basis of Renaissance symbolism, we are inclined to follow Klein's interpretation. In Renaissance texts, where the supposed problem of sloppy notation does not present itself, a letter is added to the independent term, although this seems entirely superfluous.

ἀριθμὸς, the number, is used by Diophantos to indicate the unknown. It is usually written as an abbreviation, ς, or as a letter sign closely resembling it.

"But the number that has none of these characteristics, but consists of an undetermined number of units, we call the *arithmos* and its sign is ς"[46]. The earliest known use of the symbol ς is in Papyrus Michigan 620, dating back to the first or the early second century[47].

Any declension of the word ἀριθμὸς is indicated in an exponent. For instance, ς" means ἀριθμόν. The symbol is duplicated for a plural ςς^{οι}, ςς^{ούς}, ςς^{ῶν}, ςς^{οῖς}. It is followed immediately by the number, thus ςς$^{οι}\overline{\lambda\beta}$ means 32x.

On the basis of all of the above considerations, Thomas Heath concludes that the

---

[41]H. GERSTINGER & K. VOGEL(1932), pp.14 and 22.

[42]W. SCHMIDT(1899) changed the reading οδ⋔ι'δ', $74 - \frac{1}{14}$ into ογ $\overset{ιδ'}{\overline{ιγ}}$, $73\frac{13}{14}$ (pp.156-157, l.8 and 10). The sign appears only twice, and on the same page, in *Metrika*, so it is not clear whether this is a later addition or whether other 'minus symbols' may have been resolved by scribes. F. CAJORI(1993), pp.73-74,, T. HEATH(1981), p.459, K. BARNER(2007), p.27.

[43]P. TANNERY(1912-40)III, p.160, T. HEATH(1964), p.39.

[44]J. KLEIN(1992), p.131.

[45]Diophantos I, introduction

[46]Diophantos I, introduction

[47]F.E. ROBBINS(1929), K. VOGEL(1930).

sign cannot be interpreted as an algebraic symbol[48].

To denote powers of the unknown, Diophantos uses two symbols: δ^v and κ^v, resp. the *dynamis* and the cube of the unknown. These symbols are not equivalent to the exponent, but to the power of the exponent itself. Thus δ^v does not stand for the 2 in x^2, but for the entire expression x^2. The word *dynamis*, δύναμις, has been interpreted in different ways by different translators. In colloquial Greek, it has the meaning *power*, and this is indeed the word commonly used in translations. Bailly's dictionary, for example, states: *t. d'arithm* puissance d'un nombre, *particul.* le carré[49]. However, in mathematical usage, it is always used as *a square*. The verb δύνασθαι, *dynastai*, was originally used for transformations of surfaces in the plane. Thus, a rectangle was transformed into a square of the same area (tetragonismos)[50]. The dynamis is obtained by finding or constructing the mean proportional of length and width of the rectangle. Because the transformation of a rectangle may result in a side of a square that is not measurable in length, it would appear to have been desirable to measure these sides by their squares rather than their lengths[51]. The δ^v, δύναμις, has a special place in Diophantine terminology, in the sense that, contrary to the other powers, it always refers to the square of the unknown. The square of a number is usually referred to as *tetragon*, τετράγωνον.

Higher powers can be referred to by juxtaposing these symbols. The fourth power is $\delta^v\delta$, δυναμοδύναμις. It was already used in this sense by Heron[52] and Hippolytus[53]. Although the Greek manuscript contains no powers higher than six, it is clear from the Arabic books that the symbolism is retained. In the Arab version, we encounter expressions such as 'square square square square', representing x^8, which may have been represented in Greek as $\delta^v\delta^v\delta^v\delta$. Note that this type of notation requires two symbols, δ^v and κ^v, to denote the powers. It is obvious that any number n can be written as[54] $2x + 3y$. Thus $\delta^v\kappa$ represents x^5 and $\kappa^v\kappa$ represents x^6. Also note that the notation is additional, like the exponents in our notation.

Although the origin of the terminology is not clear, it seems to have been widely used. It is, for example, also found in a surveyor's text attributed to Varro[55]. Marcus Terentio Varro (116-27 B.C.) was the editor of an encyclopedia *De Disciplinis*, a work consisting in nine volumes, the fourth of which dealt with geometry. Although the text is in Latin, the powers are referred to in transliterated Greek:

---

[48]T. HEATH(1964), pp.32-37.

[49]L. SÉCHAN & P. CHANTRAINE(1950), p.542.

[50]A. SZABÓ(1978), pp.36-55. For an opposing view, see M. CAVEING(1982), pp.1342-1362.

[51]A. SZABÓ(1978), p.103.

[52]D. SAKALIS(1972), p.43.

[53]Hippolytus, *Refutation of All Heresies*, 1.2.10. Hippolytus lived around 200. He is commonly regarded to have been the first antipope. He died in 235, reconciled with the Church.

[54]If $n = 2m$ then $x = m, y = 0$, if $n = 2m + 1 = 2(m - 1) + 3$, then $m = m - 1, y = 1$.

[55]N. BUBNOV(1963), pp.495-503.

dynamus (problem 19), *kybus* (problem 14, 22), *dynamodynamus* (problem 20), *dynamokybus* (problem 20) and *kybokybus* (problem 22). We may reasonably assume, then, that these terms had become widespread in mathematical usage by the third century.

As we have previously mentioned, nearly all Greek mathematical works have come to us through Byzantium. If we wish to gain insight into the use of mathematics, and elementary mathematics in particular, we must therefore consider surviving papyrus fragments.

In relation to the Diophantine corpus, it is interesting to look at the notation for fractions as evidenced by these fragments[56]. Essentially, the Greeks used to deal with fractions in much the same way as the Egyptians - and indeed they acknowledged this inheritance. The Egyptian technique for transforming fractions into unit fractions was still in use in the Greek era.

Although by the second century B.C., the Greeks were familiar with the superior Babylonian system, and even though they incorporated Babylonian astronomy and geometry into their mathematics, they remained faithful to the Egyptian notation. The assimilation of the Babylonian method only began in the second century B.C., when the Greeks, following the Alexandrine conquests, came into direct contact with the culture of Mesopotamia.

The work of Heron of Alexandria in particular provides an indication of which methods were applied during the Roman era. He uses both ordinary fractions and unit fractions. The unit fraction notation is encountered on a couple of occasions in *Metrika*, five times in book I of *Metrika*, just once in book II and not at all in book III, where only ordinary fractions are used.

In contrast to this sparse use of unit fractions in *Metrika*, they appear abundantly in *Geometrika* and *Stereometrika*. In some cases, problems are solved using both ordinary and unit fractions, as if Heron were trying to clarify an arithmetical procedure. In others, ordinary fractions are used, but never without a version with unit fractions. Many problems, however, are solved using unit fractions only.

In some instances, we observe a combined use: a problem formulated by means of unit fractions and solved with ordinary fractions, after which the answer is given in unit fractions, apparently because arithmetical manipulation of ordinary fractions is easier.

Unit fractions are not used in this way by Diophantos (unless subsequently omitted by copyists). Nonetheless, there is some reference to unit fractions, including those with an unknown in the denominator, about which he asserts: "and each of these fractions has the same symbol with above it the sign χ to clarify the meaning"[57]. The symbol[58] γ^{χ} or γ' represents $\frac{1}{3}$.

---

[56] On the use of fractions by the Greeks, see W. KNORR(2004) and D.H. FOWLER(2004).

[57] Diophantos, introduction. Allard's transcription (1980) makes use of an accent $'$.

[58] Obviously this is a kind of abbreviation, similar to those used by merchants for denoting multiples of χους, a measure for liquids. Multiples such as τρίχους, τετράχους or πεντάχους were occasionally written as, resp., $\chi^{\gamma}, \chi^{\delta}, \chi^{\epsilon}$. See N. KRUIT & K.A. WORP (1999).

The use of unit fractions is not generalized. Sometimes they are written as words, such as τέτρατον, one fourth (e.g. in I.22)[59], or a fraction may be represented as a number and a unit fraction, e.g. ϛγγ′μ̊γγ′ or $3\frac{1}{3}x + 3\frac{1}{3}$.

In accordance with Greek tradition, Diophantos uses a separate symbol ∠′ for $\frac{1}{2}$. For the inverses of the unknown and its square, we encounter ϛ′ (see for example III.10 and III.11) and $δ^{υ'}$ (see for example G VI.3).

We also encounter the notation with the denominator above the nominator, like $\overset{γ}{\epsilon}$ or $\overset{γ}{\dfrac{5}{3}}$, $μ̊\,\overset{γ}{ιγ}↑ϛα$ or $\dfrac{13}{3} - x$ (G IV.32). The horizontal bar is not to be regarded as a vinculum, but as a line above the letters to indicate that the expression is a number.

Occasionally, we encounter a unit fraction in this denominator above nominator notation, e.g. $\overset{φιβ′}{α}$ for $\dfrac{1}{512}$ (G IV.28).

In this semi-semantic notation, Diophantos also manipulates what we would refer to as polynomial fractions: μέσος $μ̊η$ μορίου $δ^υαϛα$ or $\dfrac{8}{x^2 + x}$ and $δ^υαϛαμ̊η$ μορίου $δ^υαϛα$ or $\dfrac{x^2 + x + 8}{x^2 + x}$ (G IV.25).

Interestingly, he considers polynomial fractions as fractions, for which he has general rules for adding and multiplying. In G IV.36, he finds the fractions $\dfrac{3x}{x - 3}$ and $\dfrac{4x}{x - 4}$.

But the product of the first and the third number is equal to $12δ^υ$ parts of $δ^υ + 12μ̊ - 7ϛ$. On the other hand, the sum of the first and the third number is equal to $7δ^υ - 24μ̊$ parts of $δ^υ + 12μ̊ - 7ϛ$, which we find in the following way. Because we have to add fractions like $3ϛ$ parts of ϛ - 3 and $4ϛ$ parts of ϛ - 4, we multiply the ϛ of the fractions with the other denominator, that is to say $3ϛ$ with the denominator of the other fraction, which is ϛ - 4, and, on the other hand, $4ϛ$ with the denominator of the other fraction, which is ϛ - 3. In this way, we find $7δ^υ - 24μ̊$ divided by the product of the denominators, that is $δ^υ + 12μ̊ - 7ϛ$.

---

[59]Tannery (1895) sometimes transcribes this as $δ°ν$.

3.5 The structure of a Diophantine problem

Just as Euclidean theorems have a specific structure, so too do Diophantine problems. In fact, in many respects the two are quite comparable.

The typical structure of a mathematical problem was described by Proklos[60]. A proposition begins with the *protasis*, which should ideally enunciate a condition and a result following from that condition. It is distinguished from the other parts by the fact that it is always general[61]. Next, in the *ekthesis,* a particular condition is set. The *diorismos* specifies a (geometrical) relation, which is what is sought. Then, in the *kataskeue*, the constituting objects of the proof or *apodeixis* are constructed. The proof is complete once the desired result stated in the *diorismos* is obtained. Finally, a conclusion or *sumperasma* is drawn. This is usually a repetition of the *protasis* with the addition of the word 'therefore'.

The structure of a Diophantine problem corresponds with this division by Proklos. Diophantos poses his problems in a general way, without numerical data (*protasis*). Solutions are provided for specific numbers (*ekthesis*), which are given at the outset. Hence there is a partial analogy to be observed with geometrical constructions, which are also posed generally, but applied to a specific figure.

Diophantos never provides a general method, though, even if a specific example so allows. During the elaboration of the example, he sometimes adds a restriction. In other cases, he does not, even if a restriction is called for. It remains unclear whether this is due to the failure to recognize this necessity or to the inability to formulate it correctly, e.g. in terms of the characterization of the numbers.

The solution to a problem begins with determining an equation. In most cases, a certain expression has to be equalled to a square or a cube (*construction*). The square or the cube is expressed in terms of an unknown in a manner that guarantees a rational solution. The latter expression is introduced into the equation, which can then be solved using Diophantos' general rules, set out in the introduction (*demonstration*). This leads to a particular solution for the problem at hand.
Usually there is no formal *conclusion*, but just a simple confirmation that the problem has been solved.

The above-described structure is characteristic of many Greek mathematical treatises. In this sense, the *Arithmetika* is a book that fits seamlessly into the

---

[60]See R. NETZ(1999b) referring to Proklos' in *Primum Euclidis Elementorum Librum Commentaria Prologus* 203.1-207.25. He also gives etymological explanations for the Greek words.

[61]This explains why some constructions are put generally, but not proved generally. On Greek methods of proof, see J. HINTIKKA & U. REMES(1974).

Greek tradition of mathematical writing, yet its content is remarkable, original and isolated within that tradition.

We can again take the famous problem as an example.

II.8 Divide a square into two squares.

protasis (enunciation)

I propose to divide 16 into two squares.

ekthesis (setting out)

I put it that the first number is δ^v, then the other is $16\overset{\circ}{\mu} - \delta^v$.

So it is necessary that $16\overset{\circ}{\mu} - \delta^v$ is a square.

diorismos (definition of goal)

Take the square of a random multiple of ς of which the square root of 16 is subtracted.

kataskeue (construction)

Take for instance $2\varsigma - 4\overset{\circ}{\mu}$, the square of which equals $4\delta^v + 16\overset{\circ}{\mu} - 16\varsigma$.

We put this equal to $16\overset{\circ}{\mu} - \delta^v$.

If we add the lacking numbers on both sides and if we subtract equals from equals, we find that $5\delta^v$ equals 16ς and $\varsigma = \dfrac{16}{5}$.

apodeixis (demonstration)

From which it follows that one of the numbers is equal to $\dfrac{256}{25}$ and the other to $\dfrac{144}{25}$.

So the sum of the numbers is $\dfrac{400}{25}$.

sumperasma (conclusion)

3.6   The *Arithmetika*

Although we do not know the origin of the *Arithmetika* and possess little or no biographical information about its author, the history of the text itself is easier to reconstruct. We shall return to this in following chapters and sections.

We have at our disposal three versions of Diophantos' text, one Arab and two Greek, each of which is based on different manuscripts. In what follows, we shall try to describe Diophantos' text as faithfully as possible and in the order that he most probably intended (see par. 3.1, p. 45).

Like the Heronian treatises, *Arithmetika* occupies a position on the interface of logistic and arithmetic. One of the characterizing features of *Arithmetika* is that Diophantos always demands that the solution should be expressible (see for example G IV.9, p. 71). Therefore, he considers whether the solution to which his parametrization leads is *expressible*, ῥητός or *not expressible*, ἄρρητος. He never uses the terms commensurable and incommensurable. Here the usage of Diophantos and Euclid diverges: whereas Diophantos treats *expressible* and *commensurable* as synonyms, Euclid differentiates between the two.

The introduction to the *Arithmetika* is addressed to Dyonisius (see p. 47). Here, Diophantos explains the nomenclature, the symbolism and some of the algorithms that are used. He starts out by defining number as being composed of units. The number of numbers that can be formed is infinite. Evidently, this refers to the Greek conception of number since Eudoxos and Euclid (see par. 1.4). Diophantos gives the nomenclature for the powers of numbers and the powers of the unknown, up to the sixth power. However:

> The number which has none of these characteristics but holds an undetermined number of units we call *arithmos* and its sign is ς.

Subsequently we are introduced to the inverses of the powers of the unknown and the products of the inverse of the unknown with the powers of the unknown. Next, he gives the product of each inverse of a power of the unknown with each power of the unknown. These products shall be required in solving the problems.

He then gives two algebraic rules:

> And next, if you find that a problem leads to an equation in which certain quantities, when equalled to one another, do not have the same coefficient, then it is necessary to subtract like from like on both sides, until only one quantity is equal to another. If by chance in one or in both sides there is a lacking [quantity], you need to add the lacking [quantity] in both sides, until on each side the existing [quantities] appear, then subtract the quantities of the same nature, until there is only one quantity of a specific nature in each of the sides.

In modern usage:

- if the equation is formed, like terms are to be added on both sides resulting in only positive terms on both sides.

- like terms on both sides cancel each other out.

For instance, in problem II.11 we find:

If we choose the square of ς - 4,	choose $y = x - 4$
the square is equal to $\delta^v + 16\overset{\circ}{\mu} - 8\varsigma$,	then $y^2 = x^2 + 16 - 8x$
which we put equal to $\delta^v + 1\overset{\circ}{\mu}$.	$x^2 + 16 - 8x = x^2 + 1$
If we add the lackings on both sides	$x^2 + 16 - 8x + 8x = x^2 + 1 + 8x$
and we deduct equals from e-quals,	$x^2 - x^2 + 16 = x^2 - x^2 + 1 + 8x$
then we see that 8ς is equal to $15\overset{\circ}{\mu}$ and $\varsigma = \dfrac{15}{8}$.	$15 = 8x$ $x = \dfrac{15}{8}$

Finally Diophantos recommends that the equation be reduced whenever possible to one power of the unknown. Not unimportantly, he announces that he intends to explain how to solve cases in which two terms are left equal to one term[62].

Book I is the starting point for an introduction to indeterminate problems. It consists mainly of linear problems, which are dealt with in Babylonian texts and abacus manuscripts alike. It is very unlikely that either the *Arithmetika* was influenced by the former or that it influenced the latter. This kind of problem is the first to be dealt with in any mathematical evolution and, more often than not, it is part of the oral tradition. The first set of problems consists of determinate problems that depend on a parameter. Of course, in the Diophantine solution, a value is given to the parameter and thus they become simple systems of linear equations.

The first problem is simple:

Divide a number into two numbers with a given difference[63].

---

[62]I.e. where more than one power of the unknown is present.

[63] $\begin{cases} x + y & = & a \\ x - y & = & b \end{cases}$

The problems up to I.25 (with the exception of problem 14, which is an indeterminate problem of the second degree[64]) are all systems of linear equations with up to six unknowns. Problems 22 to 25 are indeterminate problems. These (and problem 14) are reduced to determinate problems by giving the parameters a value. Problem I.26 is an intermediate problem between linear and quadratic:

> Given two numbers, find a number which, multiplied with the first gives a square and with the other the root of that square[65].

The first problems of the second degree are systems that would have been quite familiar by the time of Diophantos. Even our Babylonian scribes would have been able to solve them.

> Find two numbers given their sum and product.[66]

The other types of Babylonian propositions are dealt with in the next problems.

This gentle introduction is concluded with book II. Books II and III reveal *the* characteristic of the *Arithmetika*: the solution of indeterminate problems. Diophantos sets out in search of positive, expressible values of the unknown, which make the expression into a square (or other powers in the following books). In general, he is satisfied with one solution, although he is aware that more solutions exist. He always solves these problems by giving the parameter a certain value, making the problem determinate.

Book II begins with five problems that are in fact variations on problems I.31-34 (and I.14). While, in the latter, the ratio between the two unknowns is given, this is not the case in II.1-5[67].
For instance:

> I.33 Find two numbers, in a given proportion, such that the difference of their squares has a given proportion to their sum.

> II.5 Find two numbers such that the difference of their squares has a given proportion to their sum.

Problems 6 and 7 follow logically from the first five problems. In problem 6, the difference of the squares exceeds the difference by a given number, while in problem 7, it exceeds an expressible multiple of the difference by a given number.

---

[64]This problem breaks the logical order of the problems and is totally out of place in this series. We may therefore speculate that it is a later addition.

[65] $\begin{cases} ax &= \alpha^2 \\ bx &= \alpha \end{cases}$

[66] $\begin{cases} x+y &= a \\ xy &= b \end{cases}$

[67]Bearing this in mind, II.1 and I.31, II.2 and I.34, II.3 and I.14 and corollaries to I.34, II.4 and I.32, II.5 and I.33 resp. deal with the same problem.

Problem 8 is of course the most famous problem (see p. 58), although its solution was known long before Diophantos wrote his book:

> II.8 Divide a given square into two squares.

Geometrically, it can be interpreted as: given a hypotenuse, find the perpendiculars.

Problem 11 is the first in which we encounter the often-used Diophantine method of the *double equation* (see par. 3.7 esp. p. 83). With the exception of problems 17 and 18, all subsequent problems are indeterminate problems of the second degree[68].
For example:

> II.19 Find three squares, such that the difference between the largest and the middle has a given ratio to the difference of the middle and the smallest[69]

Problems II.32-35 are the extension to three unknowns of problems II.20-23 in two unknowns.
For example:

> II.20 Find two numbers such that the square of either added to the other makes a square.

> II.32 Find three numbers such that the square of any one of them added to the next number makes a square[70].

The first four problems in book III resemble the final two problems in book II. E.g.

> II.34 Find three numbers such that the square of any one, increased with the sum of all three, makes a square.

> III.2 Find three numbers such that the square of their sum, increased with any one of these numbers, makes a square.

---

[68] Problems 17 and 18 are more in line with the first 25 problems of book I. Again, we may ask ourselves whether these problems are not interpolations by later commentators. See also P. TANNERY(1895)I, p.109, note 1

[69] $\dfrac{x^2 - y^2}{y^2 - z^2} = m.$

[70] II.20 $\begin{cases} x^2 + y &= \alpha^2 \\ x + y^2 &= \beta^2 \end{cases}$ and II.32 $\begin{cases} x^2 + y &= \alpha^2 \\ y^2 + z &= \beta^2 \\ z^2 + x &= \gamma^2 \end{cases}$

Again, we are faced with a question of order. Might the latter problems actually have been the first problems of book III in the original version[71]?
Problem 10 is the first in which the method of false position is used.

> III.10 Find three numbers such that the product of any two, increased by a given number, makes a square.

Diophantos arrives at (the equivalent of) the expression $52t^2 + 12$, which must equal a square[72], with $52 = 4.13$. Neither 52 nor 12 are squares, therefore none of the Diophantine solution methods are applicable (see par. 3.7, p.81). For this reason, he seeks numbers for which the coefficient becomes a square, and these are 4 and $\frac{1}{4}$. The resulting expression is $t^2 + 12$, which must equal a square. Diophantos puts it that the square equals $(t + 3)^2$, from which $t = \frac{1}{2}$.
He could have simplified matters by choosing a square for the number that has to be added[73]. We may therefore reasonably assume he wanted to demonstrate a method that is sufficiently general to the problem with any given number.
The solutions to problems 17 and 18 are based on the application of the algebraic identity[74] $a(4a - 1) + a = (2a)^2$.
In problem 19, Diophantos notes that 65 is a sum of two squares in two ways, since 65 is the product of 13 and 5, each of which numbers is the sum of two squares. This is remarkable, because he is not dealing with rational numbers, but with integers[75]!
Problems II.20-21 are the same problems as respectively II.15 and II.14, but with a more elegant solution thanks to a more appropriate choice of parameters.

The book that follows the third book is most probably the Arabic book IV, not the Greek book IV[76]. Apparently Diophantos had conceived Book A IV as a new part of the *Arithmetika*, for he begins by stating that, in the preceding part, he has dealt with linear and quadratic problems *organized in categories which beginners can memorize*. Diophantos puts it to the reader, who cannot be but

---

[71] Tannery has argued that these four problems are additions by later commentators (P. TANNERY(1895) I, p.139. We tend to disagree. While these six problems are of the same type as those posed in book II, it is not uncommon in the *Arithmetika* for similar problems to appear in different places in the book. In most such instances, the later problems are either generalizations or an extra condition is added.

[72] The equation $52t^2 + 12 = \alpha^2$ actually has a solution for $t = 1$, for which $52.1 + 12 = 64 = 8^2$.

[73] For example, choose $a = 16$. The resulting equation then is $52t^2 + 16 = \alpha^2$. Put $\alpha = t + 4$ then $52t^2 + 16 = t^2 + 8t + 16 \Leftrightarrow t = 0 \vee t = \frac{8}{51}$.

[74] In problem 18, $b(4b - 4) - (4b - 4) = (2b - 2)^2$ is used, which, by putting $a = b - 1$, becomes the same identity after a simple manipulation.

[75] Fermat notes that, if a number is the product of two prime numbers, which can be expressed as the sum of two squares, it can be expressed as a sum of squares in two ways, and he generalizes the result. E. BRASSINE(1853), pp.65-66.

[76] The description of the Arabic books is based on J. SESIANO(1982) and R. RASHED(1984). Quotations are from Sesiano's translation, and they are followed by a reference.

Dyonisius, his addressee in the first book, that he intends to offer him 'experience and proficiency'. He announces that, in the present book, he will introduce solid numbers (products of three unknowns) either in problems, alone or in conjunction with linear (unknown of the first degree) and plane (products of two unknowns) numbers. Book A IV therefore instructs the reader on how to deal with higher powers and how to choose the power best suited for the required magnitudes. Book A IV begins with a recapitulation of the powers of the unknown, starting with the cube and up to the sixth power, their multiplication and their division. Higher powers, the eight and the ninth, are explained when they are first encountered in A IV.29.

To solve these, he introduces a third algebraic rule:

- if in each term of the equation powers of the unknown are present, the equation can be divided by the smallest power of the unknown.

Although, these books also deal with higher-degree equations, the solution method is basically the same as those in the previous books. With book IV, the reader enters the integration phase, where the pupil needs to apply the method at a higher level. It also makes clear that the Diophantine methods are not algorithms that give a solution haphazardly, but that they may be seen as general methods for solving higher-degree equations.

The first four problems of book A IV involve sums and differences of cubes resp. squares equal to squares and resp. cubes, and they all lead to a linear equation. Other problems involving a square and a cube can be reduced to one of these problems.

A IV.1 *We wish to find two square numbers the sum of which is a square number*[77].

We put x as the side of the smaller cube, so that its cube is x^3, and we put as the side of the greater cube an arbitrary number of x's, say $2x$; then, the greater cube is $8x^3$. Their sum is $9x^3$, which must equal a square. We make the side of that square any number of x's we please, say $6x$, so the square is $36x^2$. Then, since the side (of the equation) containing the x^2's is lesser in degree than the other, we divide the whole by x^2; $9x^3$ divided by x^2 gives $9x$, that is 9 roots of x^2, and the result of the division of $36x^2$ by x^2 is a number, namely 36. Thus $9x$, that is nine roots, equals 36; hence x is equal to 4. Since we assumed the side of the smaller cube to be x, the side is 4, and the smaller cube is 64; and since we assumed the side of greater cube to be $2x$, the side is 8, and the greater cube is 512. The sum of the two cubes is 576, which is a square with 24 as its side.

---

[77]Translation by J. SESIANO(1982), p.88.

Therefore we have found two cubic numbers the sum of which is a square, the lesser being 64 and the larger 512. This is what we intended to find.

Mathematically, this translates as and is generalized by:

$$x^3 + y^3 = \alpha^2.$$

Put $x = t$ and $y = pt$
from which $(1 + p^3) t^3 = \alpha^2$
putting $\alpha = qt$
we find $(1 + p^3) t^3 = q^2 t^2$

and $t = \dfrac{q^2}{1 + p^3}$ substituting this into the equation gives a solution.

Problem 14 is the first involving a system with squares and cubes, albeit a very easy one.

A IV.14 *We wish to find a number such that when we multiply it by two given numbers, one of the two [results] is a cube and the other a square*[78].

In the following problems, one of the two given expressions is a cube or a square and the other is its side. The magnitudes with which these are multiplied satisfy a certain condition, with the exception of problem 16, which is an indeterminate problem.

In problems 23-24, we find the sum and the difference of fourth powers equal to cubes. These problems can be reduced to a linear problem, as they are of the type $x^n \pm y^n = \alpha^{n\pm 1}$.

Problems that one might expect to encounter in the book but that are not included are $x^3 + y^3 = \alpha^3$ and $x^4 \pm y^4 = \alpha^2$, most probably because no solution was known; of course, today we know that no solution actually exists. Whether or not Diophantos realized intuitively or even suspected this to be the case is a matter of conjecture.

For solving problems 25 to 33, Diophantos reverts to the methods introduced in book II.

---

[78] $\begin{cases} kx = \alpha^2 \\ lx = \beta^2 \end{cases}$. Translation by J. SESIANO(1982), p.95.

For instance, in problems 32 and 33, a ninth-degree equation is solved:

> **A IV.32** *We wish to find two numbers, one cubic and the other square, such that the cube of the cube together with a given multiple of the product of the multiplication of the square by the cube is a square number*[79].
>
> $$\left(x^3\right)^3 + ax^3y^2 = \alpha^2.$$
>
> Put $x = t$ and $y = pt^3$
> from which
> $$\left(1 + ap^2\right)t^9 = \alpha^2$$
> putting $\alpha = qt^4$
> we find
> $$\left(1 + ap^2\right)t^9 = q^2t^8.$$

Problem 34 introduces systems of equations in which mixed sums of squares and cubes are equal to squares or cubes. Diophantos solves these in two ways, using the double-equation approach and another method that requires him first to solve an indeterminate equation.

Problems 36-39 can either be reduced to II.10 or they can be solved by means of the double-equation method.

In problems 43-44, we find cubes of cubes and squares of squares, but essentially these can be reduced to problems 36-39 and therefore II.10.

The first six problems of book A V are reducible to the form
$$\begin{cases} a^2x^{2j} + bx^{2j\pm1} &= \alpha^2 \\ a^2x^{2j} + cx^{2j\pm1} &= \beta^2 \end{cases}.$$
E.g.

> **A V.4** *We wish to find two numbers, one square and the other cubic, such that, when we increase the square of the square by a given multiple of the cube of the cube, the result is a square number, and when we decrease the same by another given multiple of the cube of the cube, the remainder is again a square number*[80].
>
> $$\begin{cases} y^4 + ax^9 &= \alpha^2 \\ y^4 - bx &= \beta^2 \end{cases} \text{ with } a = 5, b = 3$$
>
> Put $x = t$ and $y = pt^3$
> from which
> $$\begin{cases} p^4t^8 + at^9 &= \alpha^2 \\ p^4t^8 - bt^9 &= \beta^2 \end{cases}$$

---

[79]Translation by J. SESIANO(1982), p.109.
[80]J. SESIANO(1982), pp.128-129.

$$\Rightarrow \begin{cases} p^4 + at = \left(\dfrac{\alpha^2}{t^4}\right)^2 = \alpha'^2 \\[2mm] p^4 - bt = \left(\dfrac{\beta^2}{t^4}\right)^2 = \beta'^2 \end{cases}$$

Put $p = 2$.

Now for any square u^2:
$$\begin{cases} u^2 + 5\dfrac{u^2}{4} = \dfrac{9}{4}u^2 = \left(\dfrac{3}{2}u\right)^2 \\[3mm] u^2 - 3\dfrac{u^2}{4} = \dfrac{1}{4}u^2 = \left(\dfrac{1}{2}u\right)^2 \end{cases}$$

Taking 16 for u^2, one arrives at the solution and finds $\alpha^2 = 1536^2$ and $\beta^2 = 512^2$.

As these problems bear no relation to the other problems of book V, their origin, or at least their position in the book, has been called into question. Here we have a similar situation as with the first problems of book III, except that, in book A V, a new method is used. In these problems, Diophantos introduces a ratio.
E.g.

A V.3 *We wish to find two other numbers, one cubic and the other square, such that, when we multiply the cube by two given numbers and subtract each of the two (products) from the square, the remainder is (in both cases) a square*[81].
Let the given numbers be 12 and 7. [...]

$$\begin{cases} b^4 - la^3 = \alpha^2 \\ b^4 - ka^3 = \beta^2 \end{cases} \text{ with } l = 7, k = 12$$

Put $b = t$ and $a^3 = rt^4, \alpha^2 = m^2t^4, \beta^2 = n^2t^4$

then
$$\begin{cases} t^4 - 7rt^4 = m^2t^4 \\ t^4 - 12rt^4 = n^2t^4 \end{cases}$$

$$\Rightarrow r = \frac{1-m^2}{7} = \frac{1-n^2}{12} \Rightarrow \frac{1-n^2}{1-m^2} = \frac{12}{7} \left(= \frac{k}{l}\right).$$

Using *Elements* V.17[82]:
$$\frac{1-n^2}{m^2-n^2} = \frac{7}{5}$$

Using the preceding problem, Diophantos finds that
$$m^2 = \frac{9}{16} \text{ and } n^2 = \frac{4}{16}$$

----

[81] J. SESIANO(1982), pp.129-130.
[82] If magnitudes are proportional *componendo*, they will also be proportional *separando*. T.L. HEATH(1956)II, p.166.

Hence
$$a^3 = rt^4 = \frac{1}{16}t^4 \Rightarrow r = \frac{1}{16}$$

Put $a = \frac{1}{2}t$ then $t = 2$

Thus $a^3 = 1, b^4 = 16, \alpha^2 = 9$ and $\beta^2 = 4$.

Problems 5 and 6 have equations of the ninth degree.

A V.5 *We wish to find two numbers, one cubic, the other square, such that, when we multiply the cube of the cube by two given numbers and add each of the two (products) to the square of the square, the result is (in both cases) a square number.*

$$\begin{cases} (b^2)^2 + k(a^3)^3 & = & \alpha^2 \\ (b^2)^2 + l(a^3)^3 & = & \beta^2 \end{cases}$$

Put $a = t, b = 2t^2, k = 12$ and $l = 5$, so $\begin{cases} 16t^8 + 12t^9 & = & \alpha^2 \\ 16t^8 + 5t^9 & = & \beta^2 \end{cases}$

Divide by t^8

Now it is known that
$$\begin{cases} 16 + 12t & = & \alpha_1^2 \\ 16 + 8t & = & \beta_1^2 \\ u^2 + 12\dfrac{u^2}{4} & = & (2u)^2 \\ u^2 + 5\dfrac{u^2}{4} & = & \left(\dfrac{3}{2}u\right)^2 \end{cases}$$

Putting $u = 16$ we immediately find that $t = 4$ is a solution.

In problems 7 to 12, the identity $4(a^3 \pm b^3) = 3(a \pm b)(a \mp b)^2 + (a \pm b)^3$ lies at the heart of the solution. From these follow conditions for expressibility. However, in these cases there is also a need for a positivity condition, which is lacking.

Problems 13-16 are again constructed from an identity
viz. $(x + a)^3 + (x + b)^3 = 2x^3 + 3x^2(a + b) + 3x(a^2 + b^2) + a^3 + b^3$.

In book A VI, Diophantos continues his pedagogical approach whereby the problems presented become progressively more complex.

The first eleven problems seem to be interpolated problems, some of which correspond to problems in book A IV[83]. Other problems (12-23) are reminiscent of the problems posed in book II, but they allow a degree higher than 2.

A VI.1 *We wish to find two numbers, one cubic and the other square, having their sides in a given ratio, such that when their squares are added, the result is a square number*[84] .

---

[83] VI.1 = IV.25, VI.2 = IV.26a; VI.3 = IV.26b, VI.5 = IV.33 corollary 1a, VI.6 = IV.33 corollary 2c, VI.7 = IV.33 corollary 1c.
[84] J. SESIANO(1982), p. 139.

$$y^6 + x^4 = \alpha^2$$

Put $x = t, y = pt$ and $\alpha = qt^3$
from which $\qquad\qquad\qquad\qquad p^6 t^6 + t^4 = q^2 t^2$

The problem is thus reduced to finding a square that equals $q^2 - p^6$
(in which p is a known number).
We therefore look for two numbers for which
$$q^2 - \left(p^3\right)^2 = m^2 \Rightarrow q^2 - m^2 = \left(p^3\right)^2 .$$
which is nothing other than II.10.

Jacques Sesiano[85] calls problems 17 and 22 "so unimaginative as to be hardly
less trivial than interpolated propositions". Nothing could be further from the
truth!

A VI.17 *We wish to find three square numbers which, when added, give
a square, and such that the first of these (three square) numbers equals
the side of the second, and the second equals the side of the third*[86].

$$\begin{cases} x^2 + y^2 + z^2 &= \alpha^2 \\ x^2 &= y \\ y^2 &= z \end{cases} \quad \text{from which } x^8 + x^4 + x^2 = \alpha^2.$$

Put $\alpha = x^4 + \dfrac{1}{2}$

from which $\qquad\qquad\qquad\qquad \alpha^2 = x^8 + x^4 + \dfrac{1}{4}$

and $x^2 = \dfrac{1}{4}$ and $\alpha = \dfrac{9}{16}$

This problem is interesting, not only because of the Diophantine methods;
by putting $\alpha = yx$ in the equation, we find
$$y^2 = x^6 + x^2 + 1$$

which is a curve of genus 2. Faltings's theorem puts that this kind of curve only
has a finite number of rational points.
J.L. Whetherell succeeded in proving that the Diophantine solution, disregarding
permutations and trivial solutions, is the only solution to the equation[87]. This is
illustrative of how deceptively easy Diophantine methods can be. Diophantos, it
seems, had mastered a technique for tackling equations to such a degree that he
knew how to choose parameters that offered him the greatest likelihood of solv-
ability.

---

[85] J. SESIANO(1982), p. 259.
[86] Translation by J. SESIANO(1982), p. 149.
[87] J.L. WHETHERELL(1994), esp. p.4.

Book A VII is very short. Diophantos begins by announcing that the problems will be of the types encountered in books A IV and A V, and he explains that their purpose is to enhance the reader's "experience and skill". In other words, Diophantos intended this book to be the last in a first series, where the reader has reached the final phase of learning; it is a final repetition, intended to consolidate the acquired knowledge.

Again, the first problems, seven in this case, are interpolated.

A VII.7 *We wish to divide a square number of cubic side into three parts such that the sum of any two is a square*[88].

$$\begin{cases} y^6 & = & x_1 + x_2 + x_3 \\ z_1^2 & = & x_1 + x_2 \\ z_2^2 & = & x_2 + x_3 \\ z_3^2 & = & x_3 + x_1 \end{cases}$$

Put $y = t$ from which $t^6 = x_1 + x_2 + x_3$
Using III.6, Diophantos finds three numbers that, taken two by two, are a square *and* whose sum is also a square.
Suppose these numbers are a_1, a_2, a_3, then $a_1 + a_2 + a_3 = \alpha^2$
Now put $x_i = a_i t^4$ from which $t^6 = (a_1 + a_2 + a_3) t^4$
and $t^2 = a_1 + a_2 + a_3 = \alpha^2$
from which $t = \alpha$

In the solution to problem 7, introduces the method also to be used in problems 8-11. The approach is in fact a clever application of the method of false position applied to quadratic problems. Diophantos considers a similar problem for which a solution is known. Because the solution is determined up to a quadratic factor, the solution to the original problem is found by multiplying with the magnitudes of the intermediate problem. Generally put, if we know that $u_0^2 = \sum u_i^2$ then $a_0^2 = \sum \dfrac{u_i^2 a_0^2}{u_0^2}$.

The following problems are again quadratic problems: a square is divided into a number of parts, for which specific relations are given.

The Greek books G IV - G VI were most likely books VIII-X in the original version. Problems are dealt with in the same way, but they are of a different order. More often than not, a subproblem is dealt with separately, or another, easier, problem has to be solved first. The first attempt at solution often leads to an insolvable equation or an equation without expressible solutions. In particular, this is the case in G VI, where the solutions are perpendicular triangles with conditions on its sides and/or area.

---

[88] J. SESIANO(1982), p.160.

In book G IV, we find indeterminate equations of the third degree. Diophantos does not actually solve third-degree equations, as all his third-degree problems are reduced to the second degree, by a careful selection of the parameter.

For instance

> G IV. 1 Divide a given number into two cubes, given the sum of its sides[89]

> Divide 370 into two cubes the sum of whose sides is 10.
> I put it that the side of the first cube is equal to $\varsigma + 5$, which is half of the sum of the sides. Then the other side equals $5 - \varsigma$. From this it follows that the sum of the cubes equals $30\delta^v + 250$. Which we equal to 370, that is to say, the given number, and $\varsigma = 2$.
> If we return to the problem posed, then the side of the first cube equals 7 and that of the second cube 3, and the cubes themselves are equal to 343 and 27.

Problem 9, goes as follows:

> G IV.9 Add the same number to a cube and its side and make them the opposite[90].

It clearly demonstrates that Diophantos allows only expressible solutions. Whereas, in the other problems, the fact that the solutions are expressible is more or less a by-product of the choice of parameters, here Diophantos demands expressibility for the solution.

Diophantos arrives at the equation $35t^2 = 5$, the solution of which is non-expressible "because the ratio of the one number to the other is not the proportion of a square to a square"[91].

---

[89] $\begin{cases} x^3 + y^3 &= a \\ x + y &= b \end{cases}$.

Put $\begin{cases} x &= \dfrac{b}{2} + t \\ y &= \dfrac{b}{2} - t \end{cases}$, then $\begin{cases} x^3 &= \dfrac{b^3}{8} + 3\dfrac{b^2}{4}t + 3\dfrac{b}{2}t^2 + t^3 \\ y^3 &= \dfrac{b^3}{8} - 3\dfrac{b^2}{4}t + 3\dfrac{b}{2}t^2 - t^3 \end{cases}$,

from which $x^3 + y^3 = \dfrac{b^3}{4} + 3bt^2 = a$ and $t^2 = \dfrac{40 - b^2}{12}$.

t will be expressible if the fraction is a perfect square. Therefore the system of equations will have expressible solutions if $a = \dfrac{12bt^2 + b^3}{4}$ for some expressible t.

In Diophantos' solution, $b = 10$ and $t = 2$ giving $a = 370$.

[90] I.e. the sum of the number and the cube is a certain number, while the sum of the number and the side is the cube of that certain number, or $\begin{cases} y + x^3 &= \alpha \\ y + x &= \alpha^3 \end{cases}$.

[91] This expression was also used by other mathematicians, such as Proklos and Pappos, and seems to refer to *Elements* X.9. However, this is a source for *geometric* definitions. Therefore, Knorr, on the basis of a passage in Heron's *Definitions* argues that a similar arithmetic

Problem 19 is interesting, because here Diophantos allows the solution to be put in function of a parameter.

> G IV.19 Find three undetermined numbers, such that the product of any two, increased by unity, makes a square.

[...] We put it that the second number is ς, then the first number is $\varsigma + 2\overset{\circ}{\mu}$.

On the other hand, as the square of $2\varsigma + 1\overset{\circ}{\mu}$ equals $4\delta^{\upsilon} + 4\varsigma + 1$, if we subtract 1 in the same fashion, we are left with $4\delta^{\upsilon} + 4\varsigma$. Now the product of the second and third number has to equal $4\delta^{\upsilon} + 4\varsigma$, and the second number is ς, so the remaining number is $4\varsigma + 4\overset{\circ}{\mu}$.

In this way, we have solved the problem undetermined in such a way that the product of two random numbers increased by one makes a square and ς is any number.

Because solving in the indeterminate means making an expression for which, for any value of ς one wants in the expression, the conditions will be satisfied.

In problem 25, Diophantos introduces yet another technique: the method of limits. This allows him to find a value of the unknown for which a certain function of the unknown(s) takes on a value between two other functions of the unknown(s).

> G IV.25 Divide a given number into three numbers such that their volume makes a cube whose side equals the sum of the differences of the numbers[92].

[...]

Suppose the smallest number is ς. It then follows that the largest number is equal to $\varsigma + 1$. From this, it follows that if we divide 8 by their product, that is to say by δ^{υ}, we find 8 divided by $\delta^{\upsilon} + \varsigma$ as the middle number. Now we want this fraction to be larger than ς on the one hand and smaller than $\varsigma + 1$ on the other hand. And because the difference of these latter numbers equals 1, it follows that the difference between the first and the second number is smaller than 1, hence the second number, increased by 1, is larger than the second number. But the second number increased by 1, and simplified with $\delta^{\upsilon} + \varsigma$, becomes $\delta^{\upsilon} + \varsigma + 1$ divided by $\delta^{\upsilon} + \varsigma$, making this expression larger than $\varsigma + 1$. If we multiply everything with the denominator, then $\delta^{\upsilon} + 8$ is larger than $\kappa^{\upsilon} + 2\delta^{\upsilon} + \varsigma$. If we subtract, among other things, equals from equals, we find that 8 is larger than $\kappa^{\upsilon} + 2\delta^{\upsilon}$. [...]

---

treatise, which included definitions pertaining to incommensurables, must have existed. W. KNORR(1975), p.235.

$$92 \begin{cases} x + y + z &= a \\ xyz &= ((y-x) + (z-y) + (z-x))^3 \end{cases}$$

Problems 29 and 30 are not uninteresting for the theory of numbers, for Diophantos writes:

G IV.29 [...] Therefore we divide 13 into two squares 4 and 9 and divide these squares again into two squares notably $\frac{64}{25}$ and $\frac{36}{25}$ and in $\frac{144}{25}$ and $\frac{81}{25}$.

G IV.30 [...] therefore we have to divide 5 into four squares [...]. Now we can divide 5 into the squares $\frac{9}{25}, \frac{16}{25}, \frac{64}{25}$ and $\frac{36}{25}$[93].

Bachet[94] concludes that Diophantos must have known that any non-square number can be written as the sum of two, three or at most four squares. In both these cases, however, the number to be divided into four squares is itself conveniently the sum of two squares. And, according to II.8, any square can obviously be divided into two squares.

Problem 31 seems trivial, but its enunciation is worth noting.

G IV.31 *Divide 1 into two numbers* and add given numbers to them respectively, such that the product of the two sums makes a square.

The italicised text reveals that Diophantos had no philosophical objection to dividing unity *into numbers*. We may therefore assume that Diophantos also regarded fractional parts of the unit as numbers.

In book G V, which again deals with indeterminate problems of second and third degree, Diophantos presupposes that the reader is familiar with certain properties of numbers.
In problem 3, he refers to another book, entitled the *Porisms*:

We know from the *Porisms* that, if two numbers and their product are each increased with a given number to make squares, then these numbers are brought forth by two consecutive squares.

A porism is a statement directed at finding what is proposed. It is in this sense that Diophantos uses them, i.e. as known techniques that have been proved or explained elsewhere. He also mentions the *Porismata* in problems 5 and 16[95].

---

[93] $5 = 4 + 1$ and $1 = \frac{9}{25} + \frac{16}{25}$, $4 = \frac{144}{25} + \frac{81}{25}$.
[94] G. BACHET(1621), pp. 240-242.
[95] Porism in G V.5: If $x_1 = m^2, x_2 = (m+1)^2, x_3 = 4(m^2+m+1)$ then $x_i x_{i+1} + x_i + x_{i+1} = \alpha^2$ and $x_i x_{i+1} + x_{i+2} = \alpha^2$. Porism in G V.16 see p. 75. On porisms see par. 3.3,p. 50

Problem 9 is interesting in at least two ways. The condition which Diophantos imposes seems to have been thoroughly corrupted by commentators, who perhaps could not make sense of it.

> G V.9 Divide unity into two parts such that, if a given number is added to either part, the result is a square. The given number must not be odd [and the double of this number, inrcreased by one, must not be divisible by a prime number which, added to 1, is a quadruple].

The text between square brackets is an addition by Tannery, in an attempt to correct the text. Allard[96] interpolates as:

> It is therefore necessary that the given number should not be odd and that the double, increased by one, should not be divisible by a prime number that [if itself increased by one] is a quadruple.

The problem is equivalent to the system
$$\begin{cases} x+y = 1 & (1) \\ x+a = \alpha^2 & (2) \\ y+a = \beta^2 & (3) \end{cases}.$$

Adding (2) to (3), we find $x+a+y+a = \alpha^2 + \beta^2 \Leftrightarrow 1 + 2a = \alpha^2 + \beta^2$.
For example, consider the number 1225, which is divisible by 7, and obviously $7 + 1 = 8 = 4.2$, yet $1225 = 2.612 + 1 = 784 + 441 = 28^2 + 21^2$ and 612 is not uneven. The reason here of course being that $1225 = (5.7)^2$. Now according to Diophantos II.8, a square can always be divided into two squares.

The condition should therefore read:

> It is always necessary that the given number should not be odd and that the double of this number, increased by one, *when divided by its largest square factor*, should not be divisible by a prime number that, when added to 1, makes a quadruple.

Both interpolations are equivalent to $p = 4k - 1 = 4l + 3$, in which p is a prime number. The necessity of this condition, with the aforementioned proviso, was proved by Fermat[97]. Because the text is so corrupted, it is impossible to restore Diophantos' original condition, but it is barely conceivable that he correctly stated Fermat's conditions.
It is in this problem that he uses the method of limits for the first time (see p. 85).

Problem 10 is analogous, but more complicated, because here unity has to be divided into two parts that make squares if added to different numbers.
It is however an interesting problem, because it is the only one in which geometrical terminology is used, at least in the opening paragraphs:

---

[96] A. ALLARD(1980), pp.908-909.
[97] E. BRASSINE(1853), p. 97.

> Represent the unit by the line AB and cut it in Γ. Add to $A\Gamma$ the line $A\Delta$ with length 2 and add to ΓB the line BE; so each of them, $\Gamma\Delta$ and ΓE, represents a square.

In the closing part, the terminology is typically Diophantine:

> From this it follows that, if we deduct 2 from this square, one of the parts is $\dfrac{1438}{2809}$, making the other part $\dfrac{1371}{2809}$ and the conditions are met.

Problem 16 goes as follows:

> To find three numbers such that the cube of the sum of these three numbers, decreased with any of them, makes a cube.

To solve this problem Diophantos, once again referring to the *Porisms*, states:

> we find in the Porisms that the difference of two random cubes can be transformed into [the sum of two][98] cubes[99]

In problems 21 and 22, in which three squares are sought whose product, when respectively increased or decreased by any one of them, makes a square, the resulting sixth-degree equation is reduced to a quadratic equation. The latter is solved by invoking properties of right angled triangles[100].

---

[98]Lacuna interpolated by P. TANNERY(1895).

[99]Given a and b the equation $x^3 + y^3 = a^3 - b^3$ can be solved by putting $x = t - b$ and $y = a - kt$. Substituting these values in the given equation, we find $t^3(1 - k^3) + 3t^2(ak^2 - b) + 3t(b^2 - a^2k) = 0$. Now put $b^2 - a^2k = 0$, so $k = \dfrac{a^2}{b^2}$. We then find $t = \dfrac{3(b - ak^2)}{1 - k^2} = \dfrac{3a^3b}{a^3 + b^3}$, giving $x = \dfrac{b(2a^3 - b^3)}{a^3 + b^3}$ and $y = \dfrac{a(a^3 - 2b^3)}{a^3 + b^3}$. This solution is due to Viète.

[100]Problem 21 can be rendered as $\begin{cases} x^2y^2z^2 + x^2 &= \alpha^2 \\ x^2y^2z^2 + y^2 &= \beta^2 \\ x^2y^2z^2 + z^2 &= \gamma^2 \end{cases}$

To solve the problem put $x^2y^2z^2 = t^2$ (1), choose (a_i, b_i, c_i) with $a_i^2 + b_i^2 = c_i^2$ and put

$\begin{cases} x^2 &= \dfrac{a_1^2}{b_1^2}t^2 \\ y^2 &= \dfrac{a_2^2}{b_2^2}t^2 \\ z^2 &= \dfrac{a_3^2}{b_3^2}t^2 \end{cases}$.

The conditions of the problems are fulfilled and equation (1) remains to be solved.

Now $\dfrac{a_1^2 a_2^2 a_3^2}{b_1^2 b_2^2 b_3^2}t^6 = t^2 \Rightarrow t^2 = \dfrac{b_1 b_2 b_3}{a_1 a_2 a_3}$.

This will be fulfilled if $\prod a_i b_i = \delta^2$.

Choose a right-angled triangle (a_1, b_1, c_1) and construct two other right-angled triangles using

$a_2 = 2b_1 c_1, b_2 = c_1^2 - b_1^2 = a_1^2, c_2 = c_1^2 + b_1^2$

$a_3 = 2a_1 c_1, b_2 = c_1^2 - a_1^2 = b_1^2, c_2 = c_1^2 + a_1^2$

then $\dfrac{b_1 b_2 b_3}{a_1 a_2 a_3} = \dfrac{b_1 a_1^2 b_1^2}{a_1 . 2b_1 c_1 . 2a_1 c_1} = \dfrac{b_1^2}{4c_1^2}$.

The unity is a square and, if $\dfrac{120}{720}$ were a square, the problem would have been solved. Now, this is not the case and therefore we have to find three right-angled triangles such that the volume number that is formed by their three perpendiculars multiplied with the volume number of their bases makes a square.

Let this square have as its side the product of the sides along the right angle of one of the right-angled triangles. So, if we divide everything by the product of the sides along the right angle of the right-angled triangle that we have just described, we find the product of the sides lying along the right angle of the other triangle. So if we suppose that one of the triangles is 3, 4, 5, we must find two right-angled triangles such that the product of the sides along the right angle taken together is 12 times the product of the sides along the right angle of the other triangle; which makes the area of the one 12 times larger than the area of the other, and if it is 12 times, then three times as well[101].

Now this is easy, and the one triangle is similar to the triangle 9, 40, 41 and the other is similar to the triangle 8, 15, 17. We now have three right-angled triangles, such that we can return to our original problem and put it that the sought after squares are $\dfrac{9}{16}\delta^{\upsilon}$, $\dfrac{225}{64}\delta^{\upsilon}$ and $\dfrac{81}{1600}\delta^{\upsilon}$.

The thirtieth and final problem would appear to be entirely out of keeping with the rest of the book. Here we find not a generally stated problem between algebra and the theory of numbers, but a plain logistic problem - or perhaps not quite.

> A person who was required to do something useful for his travel companions mixed a number of measures of wine of eight drachmas with a number of measures of wine of five drachmas, and for this he paid a square that, when added to the proposed number, makes a square whose side is the number of used measures of wine. So work out how much wine of 8 drachmas he had and also tell me, my lad, how much of the other wine, of 5 drachmas, he had.

The solution is not simple by any standard, and Diophantos' choice of parameters moreover complicates matters somewhat.

---

[101] This is a very enigmatic sentence, which translates mathematically as follows:

Diophantos finds that (see previous footnote): $t^2 = \dfrac{b_1 b_2 b_3}{a_1 a_2 a_3}$, which is equivalent to $a_1 a_2 a_3 b_1 b_2 b_3 = t_1^2$.

If we put, as Diophantos does, $t_1^2 = (a_2 b_2)^2$ it follows that $a_1 a_2 a_3 b_1 b_2 b_3 = (a_2 b_2)^2$ or $a_1 a_3 b_1 b_3 = a_2 b_2$.

Choosing one triangle arbitrarily as (3, 4, 5)
we find: $12 a_3 b_3 = a_2 b_2$

if, on the other hand, we put $t_1^2 = (2 a_2 b_2)^2$, then we find, as Diophantos concludes, that $3 a_3 b_3 = a_2 b_2$

It goes along the following lines:

The proposed number, which is not specified in the problem, is chosen to be 60. The conditions which are imposed then lead to the system

$$\begin{cases} x + y &= \alpha^2 \\ \alpha^2 + 60 &= \beta^2 \\ \dfrac{x}{8} + \dfrac{y}{5} &= \beta \end{cases}$$

The second equation can be rewritten as $\alpha^2 = \beta^2 - 60$, put $\alpha = \beta - k$ then $\beta^2 - 60 = \beta^2 - 2\beta k + k^2 \Rightarrow \beta = \dfrac{k^2 + 60}{2k}$ (1).

Now from the third equation $\dfrac{y}{5} = \beta - \dfrac{x}{8}$, introducing this into the first equation, we find $x + 5\beta - \dfrac{5x}{8} = \beta^2 - 60$ or $\beta = \dfrac{1}{5}\left(\beta^2 - 60 - \dfrac{3x}{8}\right)$,

from which Diophantos deduces $\beta < \dfrac{1}{5}(\beta^2 - 60)$ (2).

On the other hand, $y = 8\beta - \dfrac{8}{5}$ and analogously $\beta > \dfrac{1}{8}(\beta^2 - 60)$ (3) from which $5\beta < \beta^2 - 60 < 8\beta$.

Now from (2):

$$\beta^2 > 5\beta + 60$$

$$\Rightarrow \beta^2 - 5\beta + \frac{25}{4} > 60 + \frac{25}{4}$$

$$\Rightarrow \left(\beta - \frac{5}{2}\right)^2 > 66\frac{1}{4}$$

$$\Rightarrow \beta > \frac{5}{2} + \sqrt{66\frac{1}{4}} \approx 10,64$$

from which Diophantos deduces that β cannot be smaller than 11. From (3) $\beta^2 - 60 < 8\beta$ he analogously deduces that $\beta < 4 + \sqrt{76} \approx 12,72$. Therefore, Diophantos does not admit numbers larger than 12. From which $11 < \beta < 12$, but we know (1) that $\beta = \dfrac{k^2 + 60}{2k}$

So $11 < \dfrac{k^2 + 60}{2k} < 12$.

this leads to the inequalities $k^2 + 60 > 22k$ (4) and $k^2 + 60 < 24k$ (5). from (4) $k > 11 + \sqrt{61} \approx 18,8$ so k has to be larger than 19. from (5) $k < 12 + \sqrt{84} \approx 21,16$ so k has to be smaller than 21. Now if $19 < k < 21$ then we can put $k = 20$ and (1) becomes $\beta = 11\frac{1}{2}$ and $\beta^2 = 132\frac{1}{4}$ and $x = 4\frac{11}{12}$ and $y = 6\frac{7}{12}$.

Book G VI consists of twenty-four problems, all dealing with relations between the sides of right-angled triangles, other than the Pythagorean theorem.

While in essence these problems are no more difficult to solve than those in the previous books, they tend to be long and tedious, and demand the reader's full attention in order for the reasoning to become clear. Attention is due in particular to the way in which Diophantos constructs his triangles, as here he applies two methods: one beginning with an odd number, constructing a Pythagorean triplet according to the Pythagorean method (see par. 1.4), the other beginning with two unknown numbers the sum of whose squares expresses the hypotenuse, while the double product and the difference of their squares expresses the other sides respectively.

In book G VI, we find a synthesis of the properties introduced in the previous books.

No copies have been found of books XI-XIII in either Greek-Byzantine or Arabic libraries. Any attempt to characterize their content would therefore be purely speculative.

3.7   The algebra of the *Arithmetika*

The question also arises how extensive Diophantos' algebraic knowledge was. Like Euclid's *Elements*, the *Arithmetika* is a tightly organized book. It deals with problems involving linear and quadratic equations and problems concerning rational perpendicular triangles. Invariably, the problem is posed in a very general way, while the solution consists in a numerical example. The values which Diophantos proposes for solving the problem are almost perfectly chosen, which gives rise to the question of whether or not he was aware of a general solution method. The answer is most probably affirmative. Moreover, he also realized that the proposed solution to indeterminate equations was not unique. In III.19, he writes "Now we have learned to divide a square into two squares, in an infinite number of ways". The way Diophantos deals with linear and quadratic equations suggests that the special cases he uses are simply archetypal examples of the solution method. However, because of his example-based approach, it is very difficult to gain insight into Diophantos' mathematical reasoning.

There is a remarkable difference in style, though, between Diophantos and those we consider to be the classical Greek mathematicians. Nowhere is this as apparent as in solutions to similar problems in Euclid and Diophantos involving the system of equations $\begin{cases} x + y = a \\ xy = b \end{cases}$ and the resulting quadratic equation.

Euclid II.5
*If a straight line be cut into equal and unequal segments, the rectangle
contained by the unequal segments of the whole together with the square
on the straight line between the points of section is equal to the square
on the half.*[102]

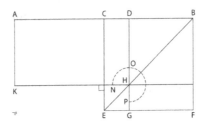

Figure 3.1

Diophantos I.27
Find two numbers such that their sum and product are given numbers.

Or better still, consider the following related problems:

Euclid (Book I Proposition 47)	*Diophantos* (Book II problem 8)
In right-angled triangles the square on the side opposite the right angle equals the sum of the squares on the sides containing the right angle.	Divide a square into two squares.
Let ABC be a right-angled triangle having the angle BAC right.	Suppose 16 is to be divided into two squares.
I say that the square on BC equals the sum of the squares on BA and AC.	I put it that the first number is δ^v, and therefore the other is $16\overset{\circ}{\mu}$ - δ^v.

---

[102]T. HEATH(1956) I, p. 382. In other words $\Box ADHK + \Box LHEG = \Box CBFE$.
In the figure, put $\mid AB \mid = a, \mid AC \mid = \mid AB \mid /2, \mid DB \mid = x$. Then $ax - x^2 =$
area gnomon $LCBFGH$. If the area of the gnomon is known, then the equation becomes
$ax - x^2 = b$.

Describe the square $BDEC$ on BC, and the squares GB and HC on BA and AC. Draw AL through A parallel to either BD or CE, and join AD and FC.

Since each of the angles BAC and BAG is right, it follows that with a straight line BA, and at the point A on it, the two straight lines AC and AG not lying on the same side make the adjacent angles equal to two right angles, therefore CA is in a straight line with AG.

For the same reason BA is also in a straight line with AH.

Since the angle DBC equals the angle FBA, for each is right, add the angle ABC to each, therefore the whole angle DBA equals the whole angle FBC.

Since DB equals BC, and FB equals BA, the two sides AB and BD equal the two sides FB and BC respectively, and the angle ABD equals the angle FBC, therefore the base AD equals the base FC, and the triangle ABD equals the triangle FBC.

Now the parallelogram BL is double the triangle ABD, for they have the same base BD and are in the same parallels BD and AL. And the square GB is double the triangle FBC, for they again have the same base FB and are in the same parallels FB and GC.

So $16\overset{\circ}{\mu}$ - δ^v must be a square.

Take the square of a random multiple of ς, from which the square root of 16 is subtracted. Take for instance 2ς - $4\overset{\circ}{\mu}$, the square of which equals $4\delta^v$ + $16\overset{\circ}{\mu}$ - 16ς.

We put this equal to $16\overset{\circ}{\mu}$ - δ^v

If we add the lacking numbers on both sides and if we subtract equals from equals, we find that $5\delta^v$ equals 16ς and $\varsigma = \dfrac{16}{5}\overset{\circ}{\mu}$.

From which it follows that one of the numbers is equal to $\dfrac{256}{25}$ and the other to $\dfrac{144}{25}$.

So the sum of the numbers is $\dfrac{400}{25}$.

In another way.

We again divide 16 into two squares.

Put it that the square root of the first number is again ς and that the square root of the second number is a random multiple of ς, decreased by the square root of the number which is to be divided, so 2ς - $4\overset{\circ}{\mu}$.

From this it follows that one square is equal to δ^v and the other to $4\delta^v$ + $16\overset{\circ}{\mu}$ - 16ς.

We want the sum of these two numbers to be equal to $16\overset{\circ}{\mu}$, so $5\delta^v$ + $16\overset{\circ}{\mu}$ - 16ς is equal to $16\overset{\circ}{\mu}$, so $\delta^v = \dfrac{16}{5}$.

Therefore the parallelogram BL also equals the square GB.
Similarly, if AE and BK are joined, the parallelogram CL can also be proved equal to the square HC. Therefore the whole square $BDEC$ equals the sum of the two squares GB and HC. And the square $BDEC$ is described on BC, and the squares GB and HC on BA and AC. Therefore the square on BC equals the sum of the squares on BA and AC.
Therefore in right-angled triangles the square on the side opposite the right angle equals the sum of the squares on the sides containing the right angle.

So the square root of the first number is $\dfrac{16}{5}$ and the number is $\dfrac{256}{25}$; the root of the second number is $\dfrac{12}{5}\overset{\circ}{\mu}$ and the number is equal to $\dfrac{144}{25}$

and the demonstration is self-evident.

Q.E.D.

The formulations and the algebraic technique proposed by Diophantos are evidently more evolved than those applied in Babylonian times. Indeterminate equations were, after all, unknown to the Babylonians. Yet the question has to be asked whether there was, in the Greek classical world, not only a geometrical tradition, but also an algebraic one, comparable to that in Babylonia. If so, then Diophantos, like Euclid in the field of geometry, was a compiler of earlier – not necessarily written – knowledge.

Let us take a look at the general methods proposed by Diophantos. He solves first-degree equations without great difficulty, but the numerical examples are always chosen in such a way that the solution is positive. He also has a remarkable sense for reducing systems of equations to a single equation. Diophantos does not however deal with indeterminate equations of the first degree[103].

For quadratic equations, Diophantos applies a standard technique, thereby indicating he knew the general algorithm. In the *Arithmetika*, he never considers more than one root, even if he seems aware that there are two positive solutions. Perhaps he was interested only in attaining *one* solution, not all solutions.
Of the indeterminate equations of the type $ax^2 + bx + c = 0$, only those are considered where a or c disappears upon application of the algorithm. The method

---

[103] On indeterminate equations of the first degree in Greek mathematics, see J. CHRISTIANIDES(1994).

Diophantos relies upon allows one to solve the general equation only if a or c or $b^2 - 4ac$ is a positive square, or if one solution is already known.

Consider the equation $ax^2 + bx + c = \alpha^2$, in which either $a = a'^2$ or $c = c'^2$. The solution is then easily found.

Suppose $a = a'^2$
The equation becomes
$$a'^2 x^2 + bx + c = \alpha^2$$
Put $\alpha = a'x + k$
and
$$a'^2 x^2 + bx + c = a'^2 x^2 + 2a'kx + k^2$$
resulting in
$$bx + c = 2a'kx + k^2$$
with solution
$$x = \frac{k^2 - c}{b - 2a'k}$$
for any value of k.
Diophantos never uses a parameter k, but always a specifically chosen value, for which the solution always happens to be positive.
Suppose $c = c'^2$
The equation becomes
$$ax^2 + bx + c'^2 = \alpha^2$$
Put $\alpha = kx + c'$
resulting in
$$ax^2 + bx = k^2 x^2 + 2c'kx$$
which, after factoring, is equivalent to the linear equations
$$x = 0 \text{ or } ax + b = k^2 x + 2c'k$$
and the solution to the latter equation is
$$x = \frac{2kc' - b}{a - k^2}$$
Again, Diophantos uses a specifically chosen value for k.

If neither $a = a'^2$ nor $c = c'^2$, Diophantos draws attention to the fact that the first choice of parametrization of α^2 may not lead to a rational solution. In IV.31, for example, he finds the expression $3x + 18 - x^2$, which must equal a square. Initially, he equals the square to $4x^2$, leading to the equation $3x + 18 = 5x^2$. He notes that, in this case, the solution is not rational[104]. He then asserts that we must find a square that, when added to 1, multiplied by 18 and then added to $2\frac{1}{4}$, is again a square.

---

[104]The solution is $x_{1,2} = \dfrac{3 \pm \sqrt{369}}{10}$.

Diophantos' expression seems enigmatic. Let us therefore transcribe it into modern notation.

We have the equation $3x + 18 - x^2 = $ a square $= \alpha^2$.

Put $\alpha = dx$.

Then

$$3x + 18 - x^2 = d^2x^2$$
$$\Leftrightarrow -(1 + d^2)x^2 + 3x + 18 = 0$$

with $D = 9 + 4.18(1 + d^2)$.

The quadratic equation will have a rational solution if D is a perfect square.

Thus $9 + 4.18(1 + d^2) = f^2$ or $\dfrac{9}{4} + 18(1 + d^2) = \left(\dfrac{f}{2}\right)^2$, which is nothing other than the condition expressed by Diophantos.

In more generally terms, this becomes:
$$ax^2 + bx + c = \alpha^2$$

with
$$\alpha^2 = d^2x^2$$

from which
$$(a - d^2)x^2 + bx + c = 0$$

with discriminant
$$D = b^2 - 4(a - d^2)c$$

Diophantos wants the latter expression to be a perfect square: $D = f^2$.

Putting
$$D' = b^2 - 4ac$$

it follows that
$$D' + 4d^2c = f^2$$

If D' is a perfect square, the previous method immediately imposes a substitution for f, with which the equation can be solved[105].

If it is not, then one of the solutions has to be known in order to be able to construct a pencil of solutions.

A very specific method applied by Diophantos is the so-called double equation. This is a system of two equations in one unknown, which must both equal a square. The solution is based on factoring of the polynomials. For instance, consider the system of linear equations:

$$\begin{cases} ax + b &= \alpha^2 \\ cx + d &= \beta^2 \end{cases}$$

By subtracting both equations, we find
$$(a - c)x + (b - d) = \alpha^2 - \beta^2$$

---

[105] In the above case: $D = 9 - 4(-1 - d^2).18 = 9 + 4.18(1 + d^2) = f^2 \Rightarrow \dfrac{9}{4} + 18(1 + d^2) = \left(\dfrac{f}{2}\right)^2$

Now suppose that
$$(a - c)x + (b - d) = p.q$$
(e.g. by considering $(a - c)$ as a common factor). We can then write
$$p.q = \left(\frac{p+q}{2}\right)^2 - \left(\frac{p-q}{2}\right)^2 = \alpha^2 - \beta^2 = (\alpha - \beta)(\alpha + \beta)$$
Diophantos immediately puts[106]
$$\alpha = \frac{p+q}{2} \text{ and } \beta = \frac{p-q}{2}$$

He also uses the method for quadratic equations. Whereas for linear equations, a solution in the rationals is always possible, this is not the case for quadratic equations.

Diophantos considers just a few examples, which always give rise to a difference that can be factored.

For example, equations of the type
$$\begin{cases} a^2x^2 + bx + c = \alpha^2 \\ a^2x^2 + b'x + c' = \beta^2 \end{cases}$$
leading to
$$(b - b')x + (c - c') = \alpha^2 - \beta^2$$
which has already been solved.
An example is G IV.23, where
$$\begin{cases} x^2 + x - 1 = \alpha^2 \\ x^2 \quad\;\; - 1 = \beta^2 \end{cases}$$
leading to $x = \alpha^2 - \beta^2$.

A second form is
$$\begin{cases} x^2 + bx + c = \alpha^2 \\ \quad\;\;\; b'x + c = \beta^2 \end{cases}$$
from which
$$x^2 + (b - b')x = \alpha^2 - \beta^2$$
with the evident resolution into factors:
$$x(x + (b - b')) = \alpha^2 - \beta^2$$

Some higher-degree equations of the type
$$\sum_{i=0}^{n} a_i x^i = y^2 (n \leqslant 6) \text{ and } \sum_{i=0}^{n} a_i x^i = y^3 (n \leqslant 3)$$
are also solved by Diophantos.

The first type of equation is solved by putting y equal to an expression that leads to the cancelling out of terms, thus resulting in an 'elementary' equation. This

---

[106]Omitting, or not recognizing, that *one* solution is given by
$$\begin{cases} p = \alpha - \beta \\ q = \alpha + \beta \end{cases}$$

approach is similar to that encountered in the solution of indeterminate quadratic equations.

As regards the second type of equation, just a few, easy-to-solve examples are found in the Greek books. In the Arabic books, on the other hand, this type of equation is dealt with extensively.

For systems of higher-order equations, he also uses the method of the double equation.

These are equations of the following type:
$$\begin{cases} ax^{2n+1} + bx^{2n} &= \alpha^2 \\ cx^{2n+1} + dx^{2n} &= \beta^2 \end{cases}$$

Put $\alpha = \alpha'x^n$ and $\beta = \beta'x^n$, which leads to the system
$$\begin{cases} ax + bx &= \alpha'^2 \\ cx + dx &= \beta'^2 \end{cases}$$

This type of system also appears in the Arabic books A IV.36-39.

In A.IV.34, a similar type of equation is solved by means of another method, which avoids the double equation. It is, however, not generally applicable.

In $\begin{cases} x^3 + y^2 &= \alpha^2 \\ x^3 - y^2 &= \beta^2 \end{cases}$

Put $x = t, y = at$ and the system becomes: $\begin{cases} t^3 + a^2t^2 &= \alpha^2 \\ t^3 - a^2t^2 &= \beta^2 \end{cases}$

Put $\alpha = vt$ and $\beta = wt$, then, on the one hand, $t = v^2 - a^2$ and, on the other, $t = w^2 + a^2$. Combining these two equations, we find that $v^2 - a^2 = w^2 + a^2$ or $v^2 - w^2 = 2a^2$, whereby the problem is reduced to finding two squares with a given difference.

More generally:
$$\begin{cases} a^2x^{2n} + by^{2n-1} &= \alpha^2 \\ a^2x^{2n} + cx^{2n-1} &= \beta^2 \end{cases}$$

Put $\alpha = \alpha'x^n, \beta = \beta'x^n$, then $y^{2n-1} = \dfrac{\alpha'^2 - a^2}{b}x^{2n} = \dfrac{\beta'^2 - a^2}{c}x^{2n}$.

Now put $y^{2n-1} = mx^{2n}$ then $\dfrac{\alpha'^2 - a^2}{\beta' - a^2} = \dfrac{b}{c}$. The problem is transformed into the known problem of finding three squares whose differences have a given ratio (II.19).

One specific method applied by Diophantos is that of limits or, as he calls it, παρρισότης or παριστότητος ἀγωγή.

In this method, two or three squares are required whose sum is a given number and each of which approximates to the same number 'as closely as possible'.

He uses the method a couple of times in G V[107]. It might also have been useful in certain problems where it is not applied[108].

In problem G V.9, Diophantos faces the problem $\begin{cases} x+y & = & 1 \quad (1) \\ x+a & = & \alpha^2 \quad (2) \\ y+a & = & \beta^2 \quad (3) \end{cases}$,

which he solves using the method of limits. As it is difficult to render the method more or less generally, we add Diophantos' numerical treatment in a footnote[109]. Obviously, from (2) + (3), it follows that $1+2a = \alpha^2+\beta^2$ and $a < \alpha^2, \beta^2 < a+1$[110].

Diophantos' method amounts to finding two numbers r, s with $a < s^2 < \gamma^2 < r^2$ such that $r^2 + s^2 = 1 + 2a$, by first determining a suitable value for γ. Since, in general, $\dfrac{1+2a}{2}$ is not a square, Diophantos seeks a square, larger than but very close to $1 + 2a$.

Now determine s such that

$$\frac{1+2a}{2} + \frac{1}{s^2} = \gamma^2$$

$$\Rightarrow \quad 2 + 4a + \frac{1}{t^2} = \gamma'^2 \quad \left(\text{having put } t = \frac{s}{4} \right)$$

$$\Rightarrow \quad (2 + 4a)t^2 + 1 = \gamma''^2$$

---

[107]G V.9-14, although in G V.10 and G V.12-14 the squares do not have to be nearly equal, but they are subject to limits.

[108]A VII.14-15. See J. SESIANO(1982), pp.274-277. It should be noted that this book precedes G V.

[109]$\begin{cases} x+y & = & 1 \\ x+6 & = & \alpha^2 \\ y+6 & = & \beta^2 \end{cases}$

$6\frac{1}{2} + \dfrac{1}{s^2} = \gamma^2$

$\Leftrightarrow \quad 26 + \dfrac{1}{t^2} = \gamma'^2$

$\Rightarrow \quad 26t^2 + 1 = \gamma''^2$

Choose $\gamma'' = 5t + 1$ and $t = 10$.

Thus $\gamma = \dfrac{51}{20}$, now $13 = 2^2 + 3^2$.

We see that $2 + \dfrac{11}{20} = \dfrac{51}{20}$ and $3 - \dfrac{9}{20} = \dfrac{51}{20}$.

Put $r = 3 - 9t$ and $s = 2 + 11t$ and suppose $r^2 + s^2 = 13$ or

$202t^2 + 13 - 10t = 13$

$\Leftrightarrow \qquad 202t^2 - 10t = 0$

$\Leftrightarrow \qquad t = \dfrac{10}{202} = \dfrac{5}{101}$

[110]J. FRIBERG(1991), p.15, calls this an additional condition. It is not, though, as it follows logically from the problem. If $x, y > 0$ then from $x + y = 1$ it follows that $x, y < 1$. From $a < x+a = \alpha^2$ and from $\alpha^2 = x + a < 1 + a$ the condition appears.

Put $\gamma" = mt + 1$ then $t = \dfrac{2m}{2 + 4a - m^2}$.

From which $2 + 4a + \dfrac{1}{t^2} = \left(\dfrac{2 + 4a + m^2}{2m}\right)^2 = \gamma'^2$

and $\gamma = \dfrac{2 + 4a + m^2}{4m}$, the value also found by the direct application of II.10 $\left[\text{if } m = \dfrac{5}{2}\right]$.

Now $\dfrac{1}{t}$ should be a small fraction.

Therefore, choose an integer value of m as close to $\sqrt{2 + 4a}$ as possible[111].
Diophantos wants to divide $1 + 2a$ into two squares, the sides [= square roots] of which are as close to γ as possible.
Suppose $r^2 + s^2 = 1 + 2a$ and $r > \gamma$, $s < \gamma$.

Suppose $r - \gamma = \dfrac{\rho}{4m}$ and $\gamma - s = \dfrac{\sigma}{4m}$

and put $r' = r - \rho x$ and $s' = s + \sigma x$.

Now suppose $(r - \rho x)^2 + (s + \sigma x)^2 = 1 + 2a$, which is a quadratic equation.

$$
\begin{aligned}
(r - \rho x)^2 + (s + \sigma x)^2 &= 1 + 2a \\
\Leftrightarrow \quad r^2 + s^2 + (2s\sigma - 2r\rho)\, x + (\rho^2 + \sigma^2)\, x^2 &= 1 + 2a \\
\Leftrightarrow \quad 1 + 2a + (2s\sigma - 2r\rho)\, x + (\rho^2 + \sigma^2)\, x^2 &= 1 + 2a \\
\Leftrightarrow \quad (2s\sigma - 2r\rho)\, x + (\rho^2 + \sigma^2)\, x^2 &= 0 \\
\Leftrightarrow \quad x = \dfrac{2(r\rho - s\sigma)}{\rho^2 + \sigma^2} \quad (\text{or } x &= 0)
\end{aligned}
$$

He uses the same method in G V.11 to find a sum of three squares.

In a modern interpretation, this procedure may be translated as follows.
Initially, $\dfrac{1 + 2a}{2}$ would be a square number.

If a square, the solution is $r = s = \sqrt{\dfrac{1 + 2a}{2}}$. If not, as in most cases, a solution is needed in which r and s are approximately equal.
In geometrical terms, Diophantos knows *one* rational point on the circle with radius $\sqrt{1 + 2a}$ and the centre $(0, 0)$, and he needs to find another rational point near the line $x = y$. Of course, this interpretation is completely anachronistic, but, as shall be demonstrated in a subsequent paragraph, there is another, historically more plausible, one.

---

[111]In G V.9 $\sqrt{2 + 4a} = \sqrt{26}$, Diophantos chooses $m = 5$, which is of course the integer value closest to $\sqrt{26}$. t will be positive for values smaller than $\sqrt{2 + 4a}$.

In conclusion, we may say that Diophantos has complete mastery of the solution methods for linear equations, for indeterminate quadratic equations and for systems of equations of the first and second degree. He also knows solution methods for some higher-degree equations. However, in some cases, his choice of parametrization seems rather fortuitous. For instance, the Arab version of problem A VI.17 goes as follows: "We wish to find three squares which, when added, give a square, and such that the first of these (three square) numbers equals the side of the second, and the second equals the side of the third."[112]. Diophantos' solution to this problem is $\left(\dfrac{1}{2}, \dfrac{9}{8}\right)$. Excepting trivial solutions and variants with minus signs, this is moreover the only solution (see par. 3.7, p. 69), but the question arises whether Diophantos was aware of this. Clearly though Diophantos' method led to *a* solution, which in his eyes was satisfactory.

3.8   Interpretations of algebra in the *Arithmetika*

Some authors, most notably I. Bashmakova and R. Rashed[113], regard Diophantos' work to be more than just technical manipulations and they draw an immediate line to analytic geometry. Bashmakova, for instance, writes:

> In his *Arithmetic* Diophantus [...] entirely solved in a purely algebraic way the problem concerning the rational points of second degree curves. In the same work he used the tangent and secant methods (again treated in an algebraic way) for the discovery of rational points on curves of third degree.

This is too much honour for Diophantos. Although the problems can be translated into contemporary mathematical terms as looking for rational points on curves, this notion is entirely absent from Diophantos' oeuvre. While he does sometimes use (ingeniously found) algorithms, he provides no justification. Still, the fact that Diophantos consistently applies them in similar fashion would appear to indicate that he realized that they were generally applicable.

Let us first apply Bashmakova's reasoning to the famous problem II.8. Subsequently, we shall consider a more plausible explanation for the origin of the Diophantine methods.

Suppose $f(x, y)$ is an irreducible polynomial of the second degree over the field $\mathbb{Q}, +, \cdot$. It is then possible to find a rational parametrization $\begin{cases} x & = & \phi(t) \\ y & = & \psi(t) \end{cases}$, which

---

[112]R. RASHED(1984) IV, p.65 and J. SESIANO(1982), p.149. The cited translation is by J. Sesiano.

[113]I. BASHMAKOVA(1981) and R. RASHED(1984). In his interpretation Rashed uses a battery of modern notions which would be in *any interpretation* alien to Diophantos or any ancient Greek mathematician.

is also the solution of the equation.

On the strength of the fact that Diophantos uses a method that is in accordance with the above method, Bashmakova concludes that he may be seen as the father of the theorem that *a rational curve of the second degree has either no rational point or is birationally equivalent to a straight line.*

In II.8, rational solutions are sought for the equation $x^2 + y^2 = a^2$ (Diophantos uses $a = 4$). Diophantos' method can be generalized by putting $y = kx - a$. The equation then becomes $(kx - a)^2 = a^2 - x^2$ and

$$\begin{cases} x & = & \dfrac{2ak}{k^2 + 1} \\ y & = & a\dfrac{k^2 - 1}{k^2 + 1} \end{cases}$$

This may be interpreted as a pencil of rational straight lines through a point $A(0, -a)$ that also intersect the circle in a point B[114].

Diophantos solves this problem numerically and finds one specific solution. However, he knows that he can find other solutions in a similar way. In III.19, he writes the following with reference to II.8: "We learned how to divide a square in two squares *in an infinite number of ways*".

In the problems following II.8, Diophantos solves quadratic equations in a similar fashion. In II.9, he implicitly states that there are other possible solutions. Here, Diophantos solves $x^2 + y^2 = a^2 + b^2$, where $a^2 + b^2 = 13(= 2^2 + 3^2)$, using the substitution

$$\begin{cases} x & = & t + 2 \\ y & = & 2t - 3 \end{cases}$$

In Bashmakova's interpretation, Diophantos discards the evident solution $(2, 3)$ and chooses instead for the solution $(2, -3)$ to construct a pencil of straight lines. All rational straight lines through A will intersect the circle in a second rational point, providing another solution to the problem.

Diophantos further indicates that there are alternative ways of solving the problem: "the second number [i.e. the coefficient of t in the expression for y] is an arbitrary multiple of t". In other words, the algorithm can be generalized by putting $y = kt - 3$.

According to Bashmakova, Diophantos' method is therefore equivalent to considering a pencil of rational straight lines through a known rational point, each of which intersects the curve in another rational point. Moreover, she assumes that Diophantos was aware of this fact (albeit in a different guise).

---

[114]I. BASHMAKOVA(1997), pp.11 & 24-25.

Another, in our opinion far more reasonable, explanation is formulated by Jöran Friberg[115], who sees a connection with the Babylonian algebraic geometrical problems. In this interpretation, Diophantos' problems are simply geometrical in nature, albeit in an algebraic rather than a geometric expression. This interpretation has the advantage that Diophantos' choice of parameters becomes very natural, without reverting to the use of rational straight lines. Let us reconsider problem II.8, but this time in the geometrical algebraic Babylonian version. It should be noted here that, in Babylonian problems, respectively the base and the large base of triangles and trapezoids are assumed to be at the top.
Consider the triangle in the figure. In anachronistic terms, the problem may be stated as follows:

> If a given symmetrical triangle, with a prescribed triangle ratio [i.e. altitude/(base/2) $= t$], which is inscribed in a circle of ratio r, in such a way that the centre of the circle divides the altitude into two parts, then calculate the base and the upper and lower part of the altitude.

The Babylonian solution (in modern notation) is:

> With reference to figure 3.2, the radius, r, is the hypotenuse of a perpendicular triangle with sides (p, q, r), in which $p = tq - r$

So
$$q^2 + (tq - r)^2 = r^2$$

and
$$(t^2 + 1)q^2 - 2tqr = 0$$

or
$$(t^2 + 1)q^2 = 2tqr$$

from which
$$q = \frac{2t}{t^2 + 1}r$$

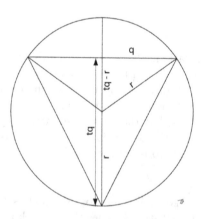

Figure 3.2

[115] J. FRIBERG(1991).

Problem II.9 can be interpreted in a similar fashion:

A symmetrical trapezoid with prescribed triangle ratio [i.e. altitude/((large base - small base)/2) = t)], and whose centre divides the altitude into two parts, is given. Suppose that the length of the small base and the lower part of the altitude are given. Then determine the upper part of the altitude and the large base.

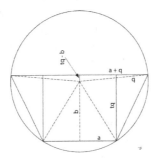

Figure 3.3

With reference to figure 3.3, we remark, that on the one hand
$$a^2 + b^2 = r^2$$
on the other hand
$$(tq - b)^2 + (a + q)^2 = r^2$$
from which
$$(tq - b)^2 + (a + q)^2 = a^2 + b^2$$
or
$$(t^2 + 1)q^2 = 2tbq - 2aq$$
from which
$$q = \frac{2tb - a}{t^2 + 1}$$

Friberg provides further convincing indications that Diophantos's method was at least similar to that used in Babylonian geometric algebra. Unlike in Bashmakova's reading, we feel that Diophantos had an algorithm that was applicable to quadratic problems and that was based on the Babylonian method or – more probably – a Greek successor[116]. In this interpretation, the disappearance of the

---

[116] It has been remarked by J. HØYRUP(2001) that the appearance of the same kind of problems does not necessarily mean there is a direct connection between the *written* sources. After all, there used to be an age-old oral tradition among craftsmen, whereby the secrets of the trade were passed down to apprentices only. Similarly in mathematical circles, the custom to keep one's methods secret prevailed at least up to the Renaissance (e.g. the feud between Cardano and Tartaglia). Diophantos appears to fit partially into this tradition. His immediate purpose was – in the practical tradition – to obtain an answer using known methods, but without indicating the origin of the method. A direct consequence is that we are unable to trace where certain problems arose or whether they arose in different places at different times. Mathematical problems appear to be very resilient in that they survive different cultures or a succession of cultures.

quadratic term follows naturally from the problem.

This is not to belittle Diophantos' merit. The problems in which he uses the algorithm are very diverse, which demonstrates his absolute mastery of the technique.
In this interpretation, however, another question comes to the fore: is Diophantos's *Arithmetika* perhaps a(n advanced) step in the de-geometrization of problems? In other words, are these in fact problems where the geometrical context has been omitted, leaving only an algebraic proposition? Or did at least some problems retain a geometrical meaning, as their formulation suggests, and were the accompanying figures omitted during copying? Alternatively, were these figures superfluous to his contemporaries?

The attractiveness of Friberg's interpretation is that it is also able to explain Diophantos' method of limits (see par. 3.7, p. 85).
Recapitulating, Diophantos needs to find two approximately equal numbers, close to $\sqrt{1 + 2a}$, in which a is a given number. Friberg argues that the Babylonians knew how to approximate square roots and that they used this knowledge to partition trapezoids into trapezoids of –nearly– equal areas.

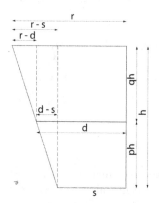

Figure 3.4

Consider a trapezoid with bases r and s and transversal d. The condition for equipartitioning[117] is $r^2 - d^2 = d^2 - s^2$ or $r^2 + s^2 = 2d^2$. If $\dfrac{r^2 + s^2}{2}$ is not a perfect square, is it possible to find near-equals r' and s' such that $r'^2 + s'^2 = r^2 + s^2$?

---

[117]Consider the similar right angled triangles in fig. 3.4 bounded on one side by the resp. dashed lines,

we find $\dfrac{r - d}{qh} = \dfrac{r - s}{h} = \dfrac{d - s}{ph}$,

This is indeed the problem that was resolved in Babylonia[118].
Rewrite the equation as $r'^2 - r^2 = s^2 - s'^2$, to make the equation determinate,
and introduce the condition that $\dfrac{r - r'}{s' - s} = \dfrac{\rho}{\sigma}$, where ρ and σ are given numbers.

The problem can now be reduced to a system of linear equations with solution:

$$\begin{cases} r' = \dfrac{(\sigma^2 - \rho^2).r + 2\rho\sigma s}{\rho^2 + \sigma^2} \\[2mm] s' = \dfrac{2\rho\sigma r - (\sigma^2 - \rho^2).s}{\rho^2 + \sigma^2} \end{cases}$$

Diophantos computed a square close to $\dfrac{13}{2} \left(= \dfrac{1 + 2a}{2} \right)$, which approximates
to the square of $\dfrac{51}{20}$.

Now $13 = 2^2 + 3^2$ and $\dfrac{3 - \dfrac{51}{20}}{\dfrac{51}{20} - 2} = \dfrac{11}{9} = \dfrac{3 - r'}{s' - 2}$, and, from the second equation,

$3 - r' = 11x$ and $s' - 2 = 9x$, which is also Diophantos' choice.

There is a second, more or less analogous interpretation in which manipulations, with a primarily geometrical meaning, are generalized and decontextualized and used in other problems. We shall give the geometrical approach, which in turn leads to the interpretation by Jean Christianides[119], who thus follows in the footsteps of Aleksander Birkenmajer[120] and Maximos Planudes.

We again take II.8 as an example. Consider the geometrical problem in figure 3.5. It is clear that $\triangle ABC$ is perpendicular. Furthermore $\triangle ABD$ and $\triangle BDC$ are similar triangles.

In these triangles, the equation $\dfrac{a + y}{x} = \dfrac{x}{a - y}$ holds.
The equality is equivalent to

$$\begin{aligned} a^2 - y^2 &= x^2 \\ \Leftrightarrow \quad x^2 + y^2 &= a^2 \end{aligned}$$

---

from which $q = \dfrac{r - d}{r - s}$ and $p = \dfrac{d - s}{r - s}$.

The trapezoid is equipartitioned if $\dfrac{r + d}{2}.qh = \dfrac{d + s}{2}.ph$.

Substituting the values for p and q we find $\dfrac{(r + d)(r - d)}{r - s} = \dfrac{(d + s)(d - s)}{r - s}$, which is the stated condition.

[118] J. FRIBERG(1991), pp.16-18.
[119] J. CHRISTIANIDES(1998).
[120] A. BIRKENMAJER (1970).

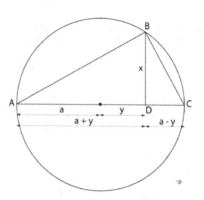

Figure 3.5

Of course, beginning with the equation, we can also retrace our steps towards the proportions. Moreover, if we put $\dfrac{a+y}{x} = k$, we find $y = kx - a$, which is in fact the substitution used by Diophantos. As x, y and a are rational, k will also be rational. Inversely, for each choice of a rational k, if x and a are rational, y will also be rational.

This technique can also be applied to II.19

$$
\begin{aligned}
x^2 + y^2 &= a^2 + b^2 \\
\Leftrightarrow \qquad x^2 - a^2 &= b^2 - y^2 \\
\Leftrightarrow \quad (x-a)(x+a) &= (b-y)(b+y) \\
\Leftrightarrow \qquad \frac{b+y}{x-a} &= \frac{a+x}{b-y}
\end{aligned}
$$

Put $t = x - a$ or $x = t + a$
and put
$$
\frac{y+b}{x-a} = k \Leftrightarrow \frac{y+b}{t} = k \Leftrightarrow y = kt - b.
$$
Again, the problem has a geometrical counterpart. It is a known property that $\triangle ABM$ and $\triangle MCD$ are similar triangles, from which the given proportion immediately follows.

Taking the equation $y^2 = a^2x^2 + bx + c$ as another example, it is clear that
$$
y^2 - a^2x^2 = bx + c
$$
from which
$$
(y - ax)(y + ax) = bx + c
$$
from which the possible proportion
$$
\frac{y - ax}{1} = \frac{bx + c}{y + ax}
$$

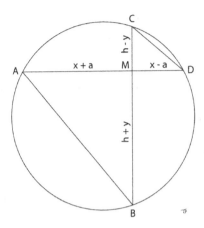

Figure 3.6

Putting the first fraction equal to k, we find $y = k - ax$, which again is the Diophantine substitution.

Christianides gives the following table with possible proportions:

Equation type	Corresponding proportion
$y^2 = a^2x^2 + bx + c$	$(y - ax) : 1 = (bx + c) : (y + ax)$
$y^2 = ax^2 + bx + c^2$	$(y - c) : x = (ax + b) : (y + c)$
$x^2 + y^2 = a^2 + b^2$	$(b + y) : (x - a) = (x + a) : (b - y)$
$y^2 = a^2x^2 + b$ and $a + b = k^2$ becomes $y^2 = a^2x^2 + k^2 - a$	$(y + k) : (x - 1) = (a(x + 1)) : (y - k)$
$y^2 = ax^2 + bx$	$y : x = (ax + b) : y$
$y^2 = ax^3 + bx^2 + cx + d^2$	$(y - d) : x = (a^3x^2 + bx + c) : (y + d)$
$y^3 = a^3x^3 + bx^2 + cx + d^3$	$(y - d) : x = (a^3x^2 + bx + c) : (y^2 + dy + d^2)$
$x(a - x) = y^3 - y$	$(y + 1) : x = (a - x) : y(y - 1)$
$y^2 = a^2x^4 + 2abx^2 + b^2 - cx^3 - dx$	$(ax^2 + b - y) : x = (cx + d) : (ax^2 + b + y)$
$y^2 = x^6 - ax^3 + bx + c^2$	$(y - x^3) : 1 = (c^2 - ax^3 + bx) : (y + x^3)$

The question then remains whether Diophantos was able to use this technique with the algorithms that were known in his time. The answer is unequivocally affirmative. He definitely knew the formulae for perfect squares, for the difference of two squares[121], and for the sum and difference of cubes. Moreover, he knew how

---

[121] See for example J. SESIANO(1999), p.30, for an early Greek application of this technique.

to factor, as becomes clear in VI.6 ("The difference becomes $y^2 - 14y$, the division gives y and $y - 14$") and VI.8 ("The difference is $14y$, the division gives $2y$ and 7").

To these three main interpretations, we may add that by Yannis Thomaides[122]. In his view, Diophantos wanted to exploit the rules he had set forth in the introduction so as to obtain a *manageable equation*. In a first step, Diophantos sets up an expression, which has to equal a square or a cube. Choosing the side of the square in a particular way leads to the cancelling out of a term on either side. This term is chosen in such a way that the equation becomes *manageable*, i.e. leading to a linear equation (see par. 3.7, p. 81). In principle, infinitely many sides can be chosen, some of which will lead to, in Diophantos' view, absurd solutions. The value for the coefficient(s) chosen by Diophantos is not random, but is the *first admissible* value.

In the case of II.8, he faces the task of equalling $16 - x^2$ to a square.

Choosing a side for this square of the expression $mx \pm 4$ guarantees that 16 is cancelled out on both sides. Which value should then be chosen for m?

In the sequence:

$$\ldots, 4x - 4, 3x - 4, 2x - 4, x - 4, x + 4, 2x + 4, 3x + 4, \ldots$$

it turns out that $2x - 4$ is [reading from right to left] the starting point of an infinity of admissible cases[123].

The problem then remains how this first admissible value is found. It is here that generality fails, as the value can only be found through trial and error[124].

Clearly, Diophantos deserves some latitude here. For one thing, he notes on several occasions that the proposed side is not the only possible one, and that therefore the solution found is not necessarily the only one either. Indeed, he does not always choose the first admissible value. For instance, in II.10 the difference between two squares has to equal a given number: $(x + m)^2 - x^2 = a$, in which $a = 60$. Obviously the *first admissible* value is $x + 1$, whereas Diophantos chooses the third admissible one, $x + 3$. In problem G V.17, the expression $9x^2 + 31 - 27x$ must equal a square. Therefore, the side can be expressed as $3x - m$.

In the sequence

$$\ldots, 3x - 7, 3x - 6, 3x - 5, \ldots, 3x - 1, 3x + 1, \ldots, 3x + 4, 3x + 5, 3x + 6, \ldots$$

the first admissible cases *are* $3x - 6$ and $3x + 5$. The expression $3x + 5$ is an isolated case, considering that, for $m < -\sqrt{31}$ and $-\frac{9}{2} < m < \sqrt{31}$, x will become negative. The only integer value between $-\sqrt{31}$ and $-\frac{9}{2}$ is $m = -5$.

---

[122]Y. THOMAIDES(2005).

[123]$mx - 4$ is always admissible, but $m = 1$ leads to the trivial solution $(4,0)$ (and $(0, 4)$ which Diophantos ignores). $mx + 4$ is never admissible for $m > 1$ because it yields a negative solution for x.

[124]Or by determining the sign of the corresponding function in m, which obviously did not lie within Diophantos' scope.

The expression which Diophantos chooses is $3x - 7$, which, if, for obvious reasons, we ignore $3x + 5$, is the second admissible. Note that the solution to the problem $(x = \dfrac{6}{5})$ makes this *admissible* expression negative! We shall return to this point in the next paragraph.

3.9 Negative numbers?

We have already mentioned Bashmakova's assertion that Diophantos works within the set of rational numbers. The question therefore arises whether Diophantos ever used negative numbers, or even understood or accepted them as a concept[125]. She uncritically accepts that he did: "Diophantos introduces negative numbers." At first glance, this would ineed appear to be the case, as he seems to give a multiplication rule for the signs.

However, matters are more complicated than that. First of all, it is worth recalling that, in modern mathematical usage, the signs + and - have two different functions. On the one hand, they are an indication of whether a number is positive or negative, while on the other, they indicate binary operators on two numbers. Moreover, the minus sign is used in the additive notation as an indication for the opposite or, in other words, the symmetrical element of a number (the appearance of which also implies the existence of a neutral element, 0). In a polynomial, the sign is therefore an operator, which means that the apparition of a negative term does not imply that the coefficient as a number is, in itself, negative.
The question ultimately revolves around the semantics of the words *leipsis* and *hyparxis*.

Leipsis, λεῖψις, seems to have appeared very late in classical Greek, meaning:
  l'action de délaisser, d'abandonner, manque, défaut[126]

According to Bailly's dictionary[127], it means: *manque, omission*
with first known use:
  DYSC Synt 78,9
  DYSC = Apollonius Dyscole d'Alexandrie du milieu du 2ème s. apr. J.C.
As far as we know, Diophantos uses the word only in a mathematical context. Possible translations are "what is missing" or "a lacking". It does not mean the *subtrahend* or *subtraction*, for which ἀφαίρεσις is used.

---

[125]See K. BARNER(2007) for a detailed analysis of this topic.
[126]C. ALEXANDRE(1888), p.844.
[127]L. SÉCHAN & P. CHANTRAINE(1950), p.1178.

The word hyparxis, ὕπαρξις, may be translated as 'existing' or 'the existing'[128]. The multiplication rule becomes: "A lacking multiplied by a lacking gives an existing and a lacking with an existing gives a lacking."

The uses of the term 'existing' seems to imply that Diophantos does not regard 'lackings' (i.e. negative numbers) to be existing entities. In other words, he does not see them as an integral part of things that exist in reality, which may explain his aversion to working with negative numbers[129].

This is apparent in V.2, which asks for the solution of a system of equations.

$$
\begin{cases}
\dfrac{x}{y} = \dfrac{y}{z} & (1) \\
x + 20 = \alpha^2 & (2) \\
y + 20 = \beta^2 & (3) \\
z + 20 = \gamma^2 & (4)
\end{cases}
$$

Choose a square such that (2) is true, e.g. $x = 16$ and put $z = t^2$.
Then, because of (1) $y^2 = 16t^2$ from which $y = 4t$, whereby (3) and (4) constitute the system

$$
\begin{cases}
4t + 20 = \beta^2 \\
t^2 + 20 = \gamma^2
\end{cases}
$$

giving $t^2 - 4t = \gamma^2 - \beta^2$.
The latter equation holds if

$$
\begin{cases}
\gamma + \beta = t \\
\gamma - \beta = t - 4
\end{cases}
$$

From which $2\beta = 4$ or $\beta^2 = 4$ and $4t + 20 = 4$.
This equation has a negative solution for t, which Diophantos calls "absurd", as 4 would then be larger than 20.

Klaus Barner[130] draws attention to the fact that, for some of Diophantos' parametrizations (= sides), he could not but have noticed that these are negative for the proposed solution. This is often the case if the quadratic polynomial has the expression $a^2x^2 + bx + c$, which has to equal a square. In such cases, Diophantos chooses a square with side $ax - m$. Here, m is a rational number, chosen in such a way that it leads to a positive solution, viz. $x = \dfrac{k^2 - c}{b + 2ak}$ (see table 3.9 on page 100).

---

[128] C. ALEXANDRE(1888), p.1467, and P. SÉCHAN & P. CHANTRAINE(1950), p.1993, translate as 'existence' and, as a neologism, also as 'realité', reality.

[129] One may wonder whether the same process is at work here as with the introduction of complex numbers. Complex numbers were already used implicitly −as square roots of negative numbers − before they were recognized as such. See also section 5.3, p. 128.

[130] K. BARNER(2007).

Looking for critical values and the change in sign of x as a function of k, we find the following possibilities[131]:

- if a, b, $c \in \mathbb{R}_0^+$ then $x > 0$ if $k \in] - \sqrt{c}, -\dfrac{b}{2a}[\cup]\sqrt{c}, +\infty[$

  or $k \in] - \dfrac{b}{2a}, -\sqrt{c}[\cup]\sqrt{c}, +\infty[$

- if $b \in \mathbb{R}_0^-$ and a, $c \in \mathbb{R}_0^+$ then $x > 0$ if $k \in] - \sqrt{c}, -\dfrac{b}{2a}[\cup]\sqrt{c}, +\infty[$

  or $k \in] - \sqrt{c}, \sqrt{c}[\cup] - \dfrac{b}{2a}, +\infty[$

- if $c \in \mathbb{R}_0^-$ then $x > 0$ if $k \in] - \dfrac{b}{2a}, +\infty[$

Obviously, we need to determine which number $\sqrt{c}$ or $-\dfrac{b}{2a}$ is the largest and then choose a number k that is larger than $\max\left\{\sqrt{c}, -\dfrac{b}{2a}\right\}$. The parametrization of the side then becomes $ax - k$. While not in line with Diophantos' reasoning, this is equivalent with determining the rational intersections of an hyperbola with a straight line parallel to the asymptote.

If the coefficient of x is not a perfect square, but the independent term is, that is $ax^2 + bx + c^2$, Diophantos chooses the expression $c - kx$, $k > 0$, making $x = 0$ or $x = \dfrac{b + 2kc}{k^2 - a}$.(see table 3.2 on the next page)
A similar determination of critical values and signs can be made for which, in some cases, we have to determine whether $\sqrt{a}$ or $-\dfrac{b}{2c}$ is larger and then choose a k that is larger still. Often, Diophantos chooses the first admissible expression.

One might wonder whether a negative side really poses a problem. If we assume that the 'side' y is 'lacking', then it would be multiplied by itself in the expression $y^2 = a^2x^2 + bx + c$. According to Diophantos, the product of two 'lackings' would give an 'existing'.

According to a second interpretation, Diophantos was aware of the negative sides, but ignored them all the same. If we calculate the left-hand side by introducing the solution for x in the right-hand side, we never even notice that y may be negative. And that was precisely Diophantos' intention: to find a value for the arithmos (x) for which the expression is a perfect square.

---

[131]The case where $a \in \mathbb{R}_0^-$ does not occur. The factorization $(y - ax)(y + ax)$ always allows a to be chosen positively.

Problem	Side	Function	Solution	$m \mapsto x(m) > 0$	Proposition
II.12	$x - 4$	$y^2 = 12 + x^2$	$x = \dfrac{1}{2}$	$]-\sqrt{12}, 0[\cup]\sqrt{12}, +\infty[$	$x - m$
II.22	$x - 2$	$y^2 = x^2 + 4x + 4$	$x = \dfrac{1}{4}$	$]-2, -\sqrt{2}[\cup]\sqrt{2}, +\infty[$	$x - m$
III.7	$x - 8$	$y^2 = x^2 + 64 - 16x$	$x = \dfrac{31}{10}$	$]-2, -\sqrt{2}[\cup]\sqrt{2}, +\infty[$	$x - m$
V.17	$3x - 7$	$y^2 = 9x^2 + 49 - 42x$	$x = \dfrac{6}{5}$	$]-\sqrt{31}, -\dfrac{9}{2}[\cup]\sqrt{31}, +\infty[$	$3x - m$

Table 3.1 *Type 1, square with side* $ax - m$

Problem	Side	Function	Solution	$m \mapsto x(m) > 0$	Proposition
IV.8	$1 - 2x$	$y^2 = 4x^2 + -4x$	$x = 7$	$]-\sqrt{3}, -\dfrac{3}{2}[\cup]\sqrt{3}, +\infty[$	$1 - mx$
IV.39	$3 - 5x$	$y^2 = 25x^2 + 9 - 30x$	$x = \dfrac{21}{11}$	$]-2, -\sqrt{3}[\cup]\sqrt{3}, +\infty[$	$3 - mx$
VI.12	$3 - 3x$	$y^2 = 9x^2 + 9 - 18x$	$x = 4$	$]-\sqrt{3}, -1[\cup]\sqrt{3}, +\infty[$	$3 - mx$

Table 3.2 *Type 2, square with side* $m - ax$

3.10 Conclusion

One may wonder whether Diophantos' *Arithmetika* is algebra or number theory. We would argue that it is neither. At least, not in the sense that we generally understand these words. Diophantos' book is concerned with algorithms, some of which may be useful to an official in executing a particular task. However strange some of these problems may seem, they are no more out of the ordinary than the problems posed to Babylonian scribes.

If one replaces Diophantos' ancient symbolism with a modern notation and reads the problems aloud, one can almost hear a teacher explaining it with the aid of a blackboard. And Diophantos would appear to have been a very accomplished teacher. Comparing his problems with pseudo-Heron's indeterminate problems makes this very clear. Pseudo-Heron uses no symbolism and performs algorithms on numbers which lead to *a* solution. Diophantos, on the other hand, does use (pseudo-)symbols and his train of reasoning is very clear, even if his choice of parameter is not. Yet the very fact that contemporary scholars seek an explanation for his choice suggests that they consider his mathematics to be advanced.

The problems which Diophantos poses are of a certain *type*. He solves them in a fashion that is general enough for the reader to follow and to grasp the underlying algorithms, not unlike a teacher who provides examples to his pupils. The reader is then free to explore similar problems with other parameters.

May the *Arithmetika* therefore be likened to a recipe book? The answer is yes, but only in the sense that Euclid's writings may. Diophantos' book is, first and foremost, a masterly encyclopedic compilation of known higher-order indeterminate problems.

The question of the origin of the problems is, by lack of comparative evidence, food for speculation. Bashmakova's and Rashed's explanations, which rely heavily on analytic geometry, may be interesting for reinterpreting Diophantine problems retrospectively, but they do not contribute to an understanding of Diophantos' train of reasoning.

Friberg, for his part, succeeded in providing a historically plausible explanation for some of the techniques Diophantos uses. Christianides complements this by reinvigorating Planudes' explanation in terms of proportion.

No matter which of these interpretations one favours –and we are strongly inclined towards the view of Christianides – they all lead to the same conclusion. Initially, some of the problems are geometrical in nature, but they are gradually decontextualized and eventually lose their relation to geometry. This is, effectively, the introduction of a kind of algebra, which can develop without geometry. Other problems are introduced that have no geometrical equivalent, yet can be solved in a similar fashion.

Both Friberg's and Christianides's explanation leave one question unanswered: was Diophantos familiar with negative numbers and did he use them? We have demonstrated that some of the sides that Diophantos chooses become negative for the solution he finds. Whether or not he was aware of this is a matter of speculation. It is our belief that he was, but that he was also unable to account for it, as he lacked the mathematical language to express himself. In this context, we recall the dual meaning of our minus sign. Negative numbers most likely posed a conceptual problem, much as imaginary numbers would a millennium later. Yet mathematicians used the latter anyway, despite an imperfect notation and a lack of understanding. In fact, negative numbers may have presented an even greater conceptual challenge. After all, once one has accepted an "unnatural" kind of number (i.e. negatives), it becomes easier to accept other, even stranger ones. Nevertheless, even if Diophantos only used "lackings" (= terms to be deducted), this is already quite a significant step toward negative numbers as such.

There is no doubt in our mind that every single problem we encounter in the *Arithmetika* had already been solved before Diophantos. His *Arithmetika* is neither algebra nor number theory, but rather an anthology of algorithmic problem-solving. This does not minimize Diophantos' (nor his predecessors') merits. In this book, we can see that mathematicians realized that what we call algebraic problems fall into categories, each with a particular method of solution that is suitable for *each* problem of this type.
Some, most even, of these problems may have originated in a practical context, such as land surveying. Others may be pure mathematical *Spielerei*. Each, however, is accompanied by a general enunciation and a model solution, leaving the reader free to explore other examples.

Summarizing, Diophantos' work seems to be situated exactly at a turning point in mathematical traditions. The *Arithmetika* on the one hand seems to be the end of the road for algorithmic algebra, which began with the Babylonian scribes, while on the other marking the beginning of abstract algebra, which would be further developed by the Arab mathematicians and would reach its full potential with the emergence of sixteenth-century symbolic algebra.

Chapter 4

Sleeping beauty in the Dark Ages

4.1   The sins of the Fathers...

Much like today's investor turns to the stockbroker to predict the performance of shares – a wildly inaccurate science at best – so the Alexandrian merchant used to call on a soothsayer to foretell his gains. Fortune-tellers appear to have been ubiquitous in Alexandria: they were of all faiths and could be found on every proverbial street corner. While most were merely concerned with making a decent living, some of Alexandria's finest minds aspired to bringing science into astrology. To be able to practise astrology on a supposedly sound basis, one needed to understand the movement of the planets and stars, which in turn required complex calculations. An astrologer could not but be a mathematician and, conversely, the best mathematicians of the era were also astrologers. In fourth-century Alexandria, as in the Roman Empire, the two words actually became synonymous[1].

One such mathematicican-cum-astrologer was Theon of Alexandria (ca. 335-405), possibly the last scholar of the Museum[2]. He wrote not only commentaries on mathematical works, but also books on divination, and is credited with such intruiging titles as *On Omens*, *The Observation of Birds* and *The Voice of Raven*[3]. Theon would appear to have been a competent teacher, but a rather unoriginal mathematician. His approach consisted in improving existing manuscripts, rather

---

[1]This is also reflected in the *Historia Augusta*, in which Hadrian is referred to as *matheseos peritus* (Ael. 3,9) and Septimus Severus as *matheseos peritissimum* (S3.,9). The latter emperor seems to have had a particular interest in the horoscopes of women of nubile age. J. STRAUB(1970), p.249. The *Historia Augusta* is a late Roman collection of biographies, in Latin, of the Roman Emperors, their junior colleagues and usurpers of the period 117 to 284. The dating of the *Historia Augusta* is uncertain.

[2]Suida θ205, translated by David Whitehead. (http://www.stoa.org/sol/)

[3]C. HAAS(1997), pp.151-152.

A. Meskens, *Travelling Mathematics - The Fate of Diophantos' Arithmetic*, Science Networks. Historical Studies 41, DOI 10.1007/978-3-0346-0643-1_4, © Springer Basel AG 2010

than simply copying them and subsequently adding emendments as was the custom. He wrote commentaries on Ptolemy's *Almagest* and on the works of Euclid, probably for the benefit of his students. As a matter of fact, some of the commentaries attributed to Theon may be notes taken down by them.

His version of Euclid's *Elements* is thought to have been a collaboration with his daughter Hypatia[4]. Together they amplified Euclid and made his work more accessible to novice mathematicians. Their text became the standard edition, superseding all previous ones and eventually driving them into oblivion. However, it is also a case in point of why Greek mathematical texts should never be taken at face value, and why one must always be on one's guards for virtually untraceable alterations. For example, at one point Theon writes:

> But that sectors in equal circles are to one another as the angles on which they stand, has been proved by me in my edition of the *Elements*, at the end of the sixth book.[5]

The proposition is found in the Euclid manuscripts as if it belonged there. Without Theon's reference, we would never have had a clue that the corollary to VI.33 is not by Euclid. This example goes to show that we can never be entirely sure how close an often-copied ancient text is to its original, including in the case of manuscripts attributed to Diophantos. In fact, Theon's daughter Hypatia also edited the *Arithmetika*, giving rise to speculation that certain problems in that text may also have been inserted by her.

It is in Theon's commentaries on Ptolemy's *Almagest* that we find the earliest reference to Diophantos:

> As Diophantos says: 'Unity is invariable and always constant. When it is multiplied by itself, it keeps the same expression.'[6]

Hypatia (ca. 350/70-415)[7] lived in Alexandria during the rise of the Eastern Roman Empire. She is described as an attractive and independent woman. Unfortunately, no works by Hypatia are known to us[8]. What little we do know about her life comes from the work of her student Synesius of Cyrene (c. 373-c.414) in the *Anthologia Graecae*, a mention in the *Suida Lexicon* and some rare references in early Christian writings[9].

---

[4]On their editions see W. KNORR(1989), p.761 ff. and A. CAMERON(1990)

[5]T.L. HEATH(1956)II, p.274. Also W. KNORR(1989), p.398.

[6]A. ROME (1931-43), p.453. See P. TANNERY (1895), pp.7-8, A. MESKENS & N. VAN DER AUWERA (2006), p.5 for comparison.

[7]Her date of birth is uncertain, but is believed to have been somewhere between 350 and 370. See M.A.B. DEAKIN(2007), pp.50-51.

[8]Translations of the most important texts relating to Hypatia are published in M.A.B. DEAKIN(2007), pp.137ff. and D. FIDELER(1993), pp.57-64. J.J. O'CONNOR & E.F. ROBERTSON(1999d and e), M. DZIELSKA(1995), N. WILSON(1983), p.42. The biography of Dzielska contains a lot of information on circles around Hypatia and the circumstances of her death, but it provides few details about her mathematics.

[9]M.A.B. DEAKIN(1994), (1995) & (2007), T. ROQUES(1995), R. HOCHE(1860), G. LUCK(1958), E. LIVREA(1997). The mystery surrounding her death has given rise to many

Philostorgius, for example, asserts that "Hypatia, the daughter of Theon, was so well educated in mathematics by her father, that she far surpassed her teacher, and especially in astronomy, and taught many others the mathematical sciences."[10]. Around 400, Hypatia became the head of the Platonic school in Alexandria. She is said to have based her teachings on the work of the Neoplatonists Plotinus[11] (204-270) and Iamblichus[12] (ca. 245-ca. 305). Founded on Plato's theory of forms, Neoplatonic thought puts forward a hierarchy of spheres of being. The highest sphere is undetermined and the source of being, while its opposite is the absolute nothing. In its Plotinean version, the Neoplatonic worldview is based on a Trinity, making it compatible with Christianity. It should therefore not come as a surprise that Christianity lay at the heart of the schism in Neoplatonic teaching and its schools. The *school of Athens* (the direct successors to Plotinus) remained anti-Christian, while the *Alexandrian school* came to embrace Christianity into its teachings. Hypatia's choice for a non-Christian Neoplatonism seems to have made her a target for Christian attacks, despite her own tolerant attitude.

From the few bits and pieces we know about her life, we are able to infer that she assisted her father in the writing of a Euclid edition and that she also produced commentaries to Diophantos' *Arithmetika* and Apollonios' *Konika*, as well as to the treatises of Ptolemy. Deakin may well be right in asserting that Hypatia was a superb compiler, editor and conservator of mathematical works, especially of what we would call handbooks[13]. She may also have been a competent mathematician in what was by any account a testing era.

All commentaries indicate that Hypatia was a charismatic teacher. It is sometimes assumed that the version of Diophantos' books we know today is based on Hypatia's edition or on one of her *Commentaries*[14]. According to this view, she only added commentaries to the first six books to have been preserved. However, it has become clear by the discovery of the Arab books that at least this latter assertion is wrong. Although it would seem that all known versions can be traced back to one archetype, there are no indications whatsoever to attribute this manuscript to Hypatia[15].

---

myths about her life, some of which have been eloquently novelized.

[10] Philostorgius, *Epitome*, 9.

[11] "Having succeeded to the school of Plato and Plotinus, she explained the principles of philosophy to her auditors, many of whom came from a distance to receive her instructions. On account of the self-possession and ease of manner, which she had acquired in consequence of the cultivation of her mind, she not unfrequently appeared in public in presence of the magistrates. Neither did she feel abashed in coming to an assembly of men. For all men on account of her extraordinary dignity and virtue admired her the more." Socrates Scholasticus, *Ecclesiastical History*, XV.

[12] J.M. RIST(1965) disagrees. He claims that philosophy in 4th and 5th-century Alexandria was essentially classical Platonist.

[13] M.A.B. DEAKIN(1994).

[14] P. TANNERY(1895)II, pp.17-18 and T. HEATH(1964), p.5.

[15] All assertions to this point have always used indirect evidence. Hypatia's edition was bound to leave traces in the form of scholia. Because this is not explicitly the case it is presumed

Another recent view, advanced by Michael Deakin[16], is that the Arabic books, rather than the known Greek books, are to be attributed to Hypatia. He, unlike Rashed (see par. 4.2, p. 110), contends that the Arabic and the Greek texts differ greatly in style. While the Greek version is sparse and to the point, the Arabic version is verbose and repetitive. Deakin notes that the Arabic problems conclude with a check and add a recapitulation of the work done. In this he sees the hand of a pedagogue and, as Hypatia is considered to have been an excellent mathematics teacher, the attribution seems logical. For teachers, he adds, performing a check is a natural thing to do, even if superfluous from a mathematical point of view. Clearly, though, both views remain speculative and neither is based on convincing evidence. Unless new clues are found, it will therefore remain an unresolved question whether or not Hypatia's reading of Diophantos has survived.

The libraries of the Museum were often the target of sectarian violence. The last of the libraries associated with the temple of Serapis, is believed to have been ravaged in 391[17] during a period of sustained violence. Despite the danger, and unlike many other philosophers, Hypatia did not flee the city. At a time when Alexandrians were torn apart by religious differences, she taught students of all persuasions.

After 412, she became the focus of a struggle between the (Christian) Church and the State over which matters are the province of which authority[18]. Although there is some debate as to who led the crowd, there is little doubt Hypatia was murdered by a Christian mob in 415[19]. With Hypatia's death, the classical era of Greek mathematics, including its Ploklean late flowering, had come to an end.

---

that these have been removed by later editors to restore the text to its original state. Because scholia are not easily recognizable if they are not indicated as such some would have remained. This would be the case for II.1 and II.2 because these seem, in the whole of the Diophantine corpus, too easy. From this it is then concluded that these are additions to make it easier for students to understand the methods which are being used. All these assertions bear a high degree of speculation in them. M.A.B. DEAKIN(1994).For a detailed analysis of the descent of Diophantos manuscripts see A. ALLARD(1982-83), see also the stemma in chapter 10.

[16]M.A.B. DEAKIN(2007), pp.98-101. The same argument was already alluded to by J. SESIANO(1982), pp.71-72.

[17]J. SCHWARTZ(1966).

[18]During the period of Hypatia's death Orestes was governor of Egypt. He was a Christian, but tolerant towards other religions. His rule was contested as of 412 by the intolerant bishop Cyrillus (the later Saint Cyrillus of Alexandria), as secular and ecclesiastical authorities fought for political dominance. These opposing views regularly gave rise to sectarian violence. Hypatia's political alliances were seen as hostile to the Church. Orestes had befriended Hypatia, and their friendship made her politically suspect. It was widely rumoured that Hypatia and her teachings were the personal and intellectual driving forces behind Orestes' political opposition to Christian authority. Her eloquence and intellectual prominence were regarded by Cyril as the seeds of destruction of the Church's authority in Alexandria.

[19]"The impious writer asserts that, during the reign of Theodosius the younger, she was torn in pieces by the Homoousian party.", Philostorgius, *Epitome*, 9. It is not clear however whether she was murdered because she was famous and happened to be in the wrong place at the wrong time, or whether it was a deliberate attempt at her life because of her alliances and philosophical views.

Baghdad, and not as one might expect Byzantium, became the new mathematical capital of the world.

4.2 From Alexandria to Baghdad

In the course of the fourth century, the great migrations from Eastern Europe began to gain momentum. This would ultimately lead to the demise of the Roman Empire by the end of the fifth century. The deposition of the Emperor in 476 is generally regarded as the moment Rome fell. The centre of power had long since shifted toward the East, where Constantinople had become the capital city. The Eastern Roman or Byzantine Empire would last until 1452. Setting aside an attempt by Justinian to restore the old Roman Empire, the history of the Byzantine Empire is one of continuous retrenchment, until only a city state remained. The largest contribution of Byzantium to science is that it preserved Greek knowledge and transmitted it to Western Europe at a crucial moment in its history.

After the fall of the Roman Empire, Europe went through an age of revolutions and upheaval, driven by migrations and dominated by a fear for Atilla the Hun. Eventually, on the ruins of ancient Rome emerged the Germanic and Vandal states.

Around the same time, Arabia also experienced some revolutions that would change the course of history. During the seventh century, Islam expanded rapidly, leading to the conversion of the Arabian peninsula, North Africa, large parts of the Middle East and parts of Asia. Although this part of the world was more or less culturally homogenized, it cannot be likened to the Roman Empire. It was more like Greece, with independent regions sharing a common culture.

Much of the Greek mathematical –and other– corpus has been lost and is known only through references in texts that have withstood the ravages of time. A number of smaller works, such as Aristarchos' *On the Sizes and Distances of the Sun and the Moon*, have been preserved because they were part of late-Roman or Byzantine curricula[20] *Syntaxis*.

In the intellectual centres at the borders of the Byzantine Empire and the Persian Empire, such as Edessa, Harrān and Ras el-Ain, translators –more often than not Christians– had been active since the fifth century. Many Greek texts were translated into Syriac, yet the subsequent influence of these centres on Arabo-Persian intellectual life, especially in Baghdad, has hardly been studied. Undoubtedly, though, these Syriac translations are a crucial link in the transmission of Greek knowledge to the Arabic civilization[21]. They were the germinators of a first renaissance of Greek thought and knowledge.

---

[20] A. MESKENS, N. VAN DER AUWERA & P. TYTGAT(2006). According to Pappos, this treatise was part of the introduction to Ptolemy's .

[21] P. BENOÏT & F. MICHEAU(1995), pp.192-202, J.L. BERGGREN (1991) & (1996), J. LAMEER(1997).

The Arab conquest swept through a large part of the Hellenized world to the foot of the Himalayan mountains. Thus, Arabia came into contact not only with Greek knowledge, but also with that of Persia and India. The most powerful impulse for translations came during the Abbasid caliphate, which had Baghdad as its capital and stretched out across the Middle East and North Africa[22]. The earliest of these translations date from the end of the eighth century, and the tradition would continue for about a hundred years. They included not only Greek works, but also Sanskrit and Pahlavi treatises.

By the eighth century, the Greek corpus had already dwindled, but it goes without saying that there were more surviving manuscripts than seven hundred years later, when Renaissance scholars returned to the writings of ancient authors. Arabic scholars sent agents from Baghdad to the Byzantine Empire to search for scientific treatises (including on astrology). Several hundreds of such texts would be translated into Arabic. Unlike in Byzantium, where knowledge was merely preserved, the Arabs built on the Greek insights, furthering the study of mathematics and contributing original new methods and theorems. A number of these Greek treatises in Arab translation would find their way to Western Europe from the twelfth century, via translation centres in Sicily and on the Iberian peninsula, where the Islamic and Christian civilizations met[23].

From the seventeenth century, European libraries began collecting Arab works, for the purpose of comparing them with mediæval copies of Greek treatises. However, the larger part of Arabic collections remained in Turkish or Arabic libraries. Conservative estimates put the number of non-inventoried manuscripts at around 200000[24]. One can only guess as to what mathematical gems might be hidden among them.

The Arab world is where two mathematical traditions merged. The first was the Indian-Persian tradition, with its focus on astronomy and algebra. Here, mathematics is seen first and foremost as a tool for solving practical problems. The other is the continuation of the Hellenistic mathematical tradition, with its emphasis on geometry and deductive reasoning. From the fusion of these two traditions emerged a third, clearly identifiably, Arab mathematical tradition.

The hegemony of the Arab world extended to the borders of India, so that Islamic mathematics soon picked up Hindu numerals, which spread to the West from the seventh century onward and are known by us today as Hindu-Arab numerals.

The first known Arab mathematician of some renown is Abū Ja'far Muhammad ibn Mūsā al-Khwārizmī (Muhammad, father of Jafar and son of Musa, from

---

[22]See G.J. TOOMER(2004) and J.P. HOGENDIJK(1996), pp.35-36.

[23]On these translations, see for example P. BENOÏT & F. MICHEAU(1995), pp.213-221, P. LORCH(2001).

[24]G. TOOMER(2004) citing H. Ritter (1953), who estimated that the libraries in Istambul alone hold some 124000 manuscripts.

Khwarizim)[25], who was probably born ca. 780 in the Khwarizim region, east of the Caspian Sea. Around 820, when he had already earned himself a reputation as a mathematician in Merv, one of the capitals of the Abbasid Caliphate, he was invited to Baghdad by the Caliph. He wrote at least two mathematical treatises *Ḥisāb 'al-jabr w'al muqābala* (Calculation of Restoration and Subtraction) and *Algoritmi de numero indorum* (Calculation with Indian numerals). Ironically, the latter treatise is known only in a Latin translation. The first book, which is usually referred to as *Algebra*, deals with the different kinds of linear and quadratic equations: $ax = b, ax^2 = bx, ax^2 = b, ax^2 + bx = c, ax^2 + b = cx, ax^2 = bx + c$. The application of these expressions avoids the use of negative numbers.

Although he was able to solve these equations algebraically, they were explained geometrically. Note that we have to understand algebra as manipulating determinate equations, not, as in Diophantos, indeterminate equations[26].

Interestingly, the Arabs were the first to solve third-degree equations. This feat was first achieved by Tābit ibn Qorra (836-901) and Al-Hasan ibn al-Haitham (better known in the West as Alhazen; ca. 965-1039). Omar Khayyām (ca. 1050-1123), who was more renowned as a poet than as a mathematician, developed a solution method based on conic sections.

The earliest indeterminate equations in Arab mathematics are found in the work of Abū Kāmil (ca. 880), about whom we possess no biographical details[27]. There is nothing to suggest that he was aware of Diophantos' book[28], but he does seem to have had access to Greek sources. He may be regarded as a direct successor to al-Khwārizmī and as a link to al-Karaji. His significance to Europe lies in the fact that his work helped lay the foundation for the writings of Fibonacci, so that, indirectly, he had a profound influence on the introduction of algebra to the continent.

Abū Kāmil's *Book about Algebra* consists of three parts: the solution of quadratic equations, the applications to algebra for constructing the pentagon and the decagon, and Diophantine equations. There is only one known Arabic manuscript copy, dating from 1253[29]. There are also known to be Hebrew and Latin translations, which however lack the part on indeterminate equations[30].

Abū Kāmil uses higher powers of the unknown in a similar fashion as Diophantos: x^6 is, for example, regarded as 'cube cube', analogous to $\kappa^v \kappa$, and x^8 is seen as 'square square square square', analogous to $\delta^v \delta^v \delta^v \delta$. Although Abū Kāmil solves indeterminate equations of the type encountered in Diophantos, he uses methods

---

[25]J.J. O'CONNOR & E.F. ROBERTSON(1999f), J.A. OAKS & H.M. ALKHATEEB(2005), pp.401-402 and (2007).
[26]See S. GANDZ(1936) and R. RASHED(1994b), pp.8-21.
[27]See J.J. O'CONNOR & E.F. ROBERTSON(1999a).
[28]J. SESIANO(1977).
[29]See M. LEVEY(1966) and J. SESIANO(1977).
[30]See R. LORCH(1993) and M. LEVEY(1966).

unknown to the latter and his treatment is less systematic[31]. Interestingly, Abū Kāmil mentions that these questions circulated among the mathematicians of his age, indicating the interest of the Arabs for the solution of indeterminate equations[32].

Not surprisingly, then, this was the age when Diophantos was translated into Arabic

The Arab version was written by Qusṭā ibn Lūqā al-Ba'labakkī , who lived in the second half of the ninth century. He was a Greek Christian and a native of Baalbek (Heliopolis)[33]. He was invited to Baghdad to work as a translator. At least from 866 onwards, he is known to have made numerous translations of Greek mathematical works and he also seems to have written treatises on mathematics and medicine[34]. Apparently he left for Armenia some time before 890. The translation of the *Arithmetika* must therefore have been produced between 860 and 890.

The manuscript itself dates back to 1198. It was produced somewhere in Iran by two copyists, the first of whom did not progress beyond the first seven folios (recto and verso). As the Arabic text has already been sufficiently analyzed philologically elsewhere, we restrict ourselves to the mathematical aspects.

The translator did not transcribe the figures into contemporary Hindu-Arab numerals, but spelt them out as words. The unknown ς is referred to with a word which may be translated as 'the case' or 'the thing'.

The abbreviations that Diophantos uses for composite expressions are also spelt out in words. $\delta^v\overline{\delta}\varsigma\overline{\beta}\overline{\mu\kappa\epsilon}$ is rendered as 'four squares plus two things and twenty five units'. The fact that Qusṭā remained faithful to the original text is demonstrated indirectly by Rashed. Qusṭā also translated the first three books of the *Arithmetika*. These have been lost, but parts of the translation appear in other mathematical treatises, including *al-Bāhir* by Al-Samaw'al ibn Yahyā (ca. 1180), who cites two examples that, according to Rashed, are Qusṭā's translation of I.16 and I.26[35]. *Al-Bāhir* is a comprehensive commentary on the works of other authors. At the end of his treatise, the author also refers to Diophantos.

Around the time that Diophantos was translated into Arabic, a number theory school emerged[36]. This school did not seek solutions to an equation within

---

[31]H. SUTER(1900)(1902), M. LEVEY(1966).

[32]A. ANBOUBA(1978), pp.84-85.

[33]There has been some debate on the first name Qusṭā. According to R. RASHED(1984), p. XVII it is the Arabic transcription of the Latin name Constans (through Greek Kostas). According to J. SESIANO(1982), however, it is the transcription of Greek Kostas, a pet name for Constantine.

[34]See for example L. AMBJÖRG(2000).

[35]R. RASHED(1984). Rashed uncritically accepts as self-evident that Al-Samaw'al uses Qusṭā's translation: 'As-Samaw'al donne deux exemples empruntés à Diophante et *donc nécessairement* à la traduction d'Ibn Lūqā' (our emphasis).

[36]R. RASHED(1994b), pp. 35 and 205-210 on Diophantos' influence, and pp. 210-331 on the

the rational numbers, but within the positive integer numbers. The problems are related, because rational solutions to an equation are equivalent to the integer solutions of the homogenized equation[37]. Hence, it goes without saying that Diophantos became a welcome source of inspiration. It is in this context that Arab scholars, for the first time in the history of mathematics, used the concept of 'insolvability'.

Not long after, more or less contemporary with the translation of Diophantos into Arabic, two authors came to the fore who show an affinity with Diophantos. The first is Abū Ja'far al-Khāzin (ca. 940), who, among other things, wrote on the solution of the equation $x^2 \pm k = a^2$.
The family of al-Khāzin is believed to have hailed from the Southern Arabian Saba[38], but al-Khāzin himself would appear to have been from Khurasan, in eastern Iran. He came to the court at Rayy (south of present-day Tehran) under Caliph Adud ad Dawlah, a ruler of the Buyid dynasty, whose court was at Baghdad. The vizier of Rayy charged him in 959/60 to calculate the angle between the equator and ecliptic planes.
Al-Khāzin's number theory work is based on the work of the mathematician al-Khujandi, a contemporary at the Rayy observatory. Al-Khujandi claimed he had succeeded in proving that $x^3 + y^3 = z^3$ has no solutions in the natural numbers (which is Fermat's theorem for $n = 3$); al-Khāzin, on the other hand, maintained his proof was flawed[39].
This latter claim is the direct inducement for a correspondence between al-Khāzin and Arab mathematicians. One of the problems he treats is $\begin{cases} x^2 + a = \alpha^2 \\ x^2 - a = \beta^2 \end{cases}$.
al-Khāzin shows that the existence of numbers x, α, β with these properties is equivalent to finding numbers u and v with $a = 2uv$ and $u^2 + v^2 = \gamma^2$. The smallest number with this property is $a = 24$, which results in $5^2 + 24 = 7^2$ and $5^2 - 24 = 1^2$. The other integers, which are solutions, are multiples of 24. A similar reasoning would lead Fibonacci to define congruent numbers.

Referring to Diophantos, he also solved the equations[40] $x^2 + (y^2)^2 = z^2$ and $x^2 + y^2 = (z^2)^2$.

---

contributions of Arab scholars to number theory.
   [37]It is clear that, if (x_1, y_1, z_1) is a solution to the homogeneous equation $P(x, y, z) = 0$, then $\left(\dfrac{x_1}{z_1}, \dfrac{x_1}{z_1}, 1 \right)$ is a solution of $P(x, y, 1) = 0$.
   [38]Better known as Sheba, from the story of King Solomon and the queen of Sheba. See J.J. O'CONNOR & E.F. ROBERTSON(1999b).
   [39]R. RASHED(1994b), p.231
   [40]A. ANBOUBA(1979), pp. 134-139.

The Persian Abū'l-Wafā' al-Buzjani (940-997/8) commented on at least part of Diophantos' book[41]. He worked at the court of Caliph Adud ad-Dawlah and his son Sharaf ad-Dawlah in Baghdad. He was involved in erecting the city's short-lived observatory. He also produced translations of and annotations to the works of Euclid, Diophantos and al-Khwārizmī, which have probably all been lost. In Abū'l-Wafā's work, we find negative numbers, representing a debt. He provides some rules for calculating with such numbers. Furthermore, he did some astronomical work and drew up trigonometric tables.

It is thanks to the treatises of al-Karajī that we can be certain Diophantos' books I-III were known to the Arabs[42]. Al-Karajī lived at the end of the tenth and the beginning of the eleventh centuries. At a very young age, he left his mountainous homeland to live in Baghdad. It was here that he wrote his most important works: al-Fakhrī, Badī' and Kāfī.
In the book entitled al-Fakhrī, Al-Karajī copied large parts of Arithmetika: nearly half of book I, the larger part of book II, the whole of book III with exception of three problems, and nearly all of book IV. The work is an attempt to generalize Diophantos' method[43].
In his book Badī' al-Karajī returns to the subject of indeterminate equations. In this work, however, he does not copy Diophantos, but rather gives his own introduction to the first books of the Arithmetika. Although al-Karajī often cites the problems put forward by Diophantos, he never refers explicitly to him[44].

Al-Karajī's most important contribution to mathematics is that he, using al-Khwārizmī's and Diophantos' methods, advanced algebra by "treating the unknown in the same fashion as known quantities". In this manner he reaches the equivalent of $m, n \in \mathbb{Z} : x^m . x^n = x^{m+n}$. In his work, we also find a predecessor to the method of mathematical induction and a precursor of a table with binomial coefficients (Pascal's triangle).
In contrast to the Byzantine copies, these Arab versions of Diophantos unfortunately did not impact on the rebirth of algebra in the Renaissance in Italy or Europe. Nonetheless, as we intend to demonstrate, in Fibonacci we may recognize a link between the number theory work of the Arabs and Western European mathematicians.

---

[41]See J.J. O'CONNOR & E.F. ROBERTSON(1999c), A. ANBOUBA(1979).

[42]J. SESIANO(1982), pp.10-11, R. RASHED(1994a) and(1994b), pp.22-33.

[43]R. RASHED(1994b), p.29.

[44]On Badī' see J. SESIANO(1977).

4.3 The Byzantine connection

The name Byzantine Empire derives from the politico-administrative centre of Byzantium, which was however called Constantinople from 330 onward. Territorially, it corresponds with the Eastern Roman Empire, which was created under the administrative division in 395 of the Roman Empire in a western and an eastern part. The Eastern Roman Empire comprised the Balkans, Asia Minor, the Middle East and Egypt[45].

With the fall of the Western Roman Empire in 476, it also gained formal independence. By the seventh century, the Empire had established its own Greek identity, separating it from its Roman roots. Up until the eleventh or twelfth century, it remained a stable state, consisting of Eastern Greece and Asia Minor.

Despite social and economic decline in the course of the twelfth century, culturally speaking, the Byzantine Empire flourished once more in a territorially stable setting. Subsequently, the Empire began to crumble and alliances with the West did not work out as planned. The Crusades were disastrous for the Byzantine Empire, leading to much internal discord and the instauration of a feudal Latin Empire (1204). After the reconquest of Constantinople by Emperor Michael Palaeologus (1259/61-1282), the Latin Empire collapsed. Once again, the Byzantine Empire flourished culturally, but financially it had been weakened beyond repair. As the fifteenth century drew on, it shrunk until no more than a city state remained. On 29 May 1453, Constantinople was conquered by the Turks, dealing the deathblow to the thousand-year-old Byzantine Empire.

How and when Diophantos began to be studied in the Byzantine Empire is as much an enigma as everything else about him. We possess merely fragmentary information about intellectual life in the Empire, and even fewer details about scientific activities.

In a twelfth-century biography of Joannes Damascenus (674/5-749), Joannes Hierosolymitanus[46] refers to eighth-century scholars who studied, among other treatises, the work of Diophantos[47].

According to this testimony, Damascenus, under the tutelage of Cosmas, read the *Quadrivium* (arithmetic, music, geometry and astronomy), in which he 'was as diligent in the theory of proportions as Pythagoras and Diophantos'[48].

As there are four centuries between the biographer and his subject, it is not unthinkable that he used a contemporary example to evoke an image of erudition for his readers. More importantly, however, he relates Diophantos to the theory of proportions (see higher par. 3.8, p. 93).

It is not uncommon for Diophantos to be seen as an ancient authority in the Byzantine Empire, which sometimes leads to the erroneous attribution to him of

---

[45]On the history of the Byzantine Empire, see for example M. ANGOLD (1997).
[46]Twelfth century hagiographer and patriarch of Jerusalem.
[47]J. CHRISTIANIDES(2002), pp.153-154, P. TANNERY(1895)II, p.36.
[48]J. CHRISTIANIDES(2002), pp.153-154, P. TANNERY(1895)II, p.36.

ancient treatises or, worse still, of more recent compilations. We know of at least
three Byzantine fragments that have been wrongly attributed to Diophantos[49].

The first is *Excerpts from Diophantos' Arithmetic*[50], which was very clearly written long after Diophantos' death, considering that it uses Hindu-Arab numerals
when dealing with square roots. The manuscript is believed to have been written
around 1303.

In a second fragment, which is sometimes also attributed to Pappos, the sexagesimal system of astronomers is used for multiplication and division[51]. It was
collated with Ptolemy's *Prologomena* and is now thought to have been written in
the sixteenth century by Johannes a Mauro.

A third fragment consists of four chapters: *The plane geometry of Diophantos*,
The methods of polygons, *Treatise on the general method of polygons* and *On the
cylinder*. It is a compilation of geometric and stereometric theorems chosen from
the work of Heron of Alexandria. This volume was written in the sixteenth century[52].

The first Byzantine whom we know definitely studied the *Arithmetika* is
Michael Psellos (1018-ca.1078). As we have no positive proof that Diophantos was
also studied prior the eleventh century, we shall focus on him.

Psellos' Christian name was Constantine, which he changed to Michael upon entering a convent[53]. For the most part of his career, he was in the service of the
Emperor. According to some sources, he was, for a considerable time, the man behind the scenes, as it were. As a professor of philosophy, he gained a certain fame
that attracted many students, including Arabs. He wrote a considerable number of
treatises, on history among other things, as well as numerous funeral orations. In
his research, he adopted an unbiased attitude toward classical and non-Christian
writers, much to the dismay of the Church authorities. He came under repeated at-
tack and was required on several occasions to prove his allegiance to the Orthodox
faith and Church.

Psellos studied areas such as gnosticism, astrology, magic and alchemy, but
also classical philosophers such as Plato and Aristotle. In one of his texts, we find
references to some of the courses he taught: Aristotle, astronomy, geometry, arithmetic, and optics (esp. properties of mirrors).

In his lectures, he used the models described in Heron's *Pneumatika*, and he is
also said to have conducted classroom experiments.

In one of his letters, Psellos cites almost literally from Diophantos' book[54]. Ac-

---

[49]P. VER EECKE(1926), pp. LVIII-LIX.

[50]P. TANNERY(1895)II, p.3, text of Bibliothèque nationale, Suppl. Gr. 387, and p. III-IV. The
manuscript which also contains Heron's *Metrika* and *Stereometrika* was edited and published by
F. HULTSCH(1864).

[51]P. TANNERY(1895)II, pp.3-15, text of Bibliothèque nationale 453, and pp. IV-V.

[52]P. TANNERY(1895)II, pp.3-15, text of Bibliothèque nationale 2448, and pp. V-VI.

[53]About Psellos see N. WILSON(1983), pp.156-166. The change of name is notable in the sense
that it was customary for one's convent name to have the same initials as one's given name.

[54]P. TANNERY (s.d.)

cording to Tannery, Psellos owned a version with scholia deriving from the work of Anatolios.

Diophantos was not forgotten in the next century either: Nicephorus Blemmydes (ca. 1197 - ca. 1272) notes in his autobiography that he travelled to Skamandros to the teacher and hermit Prodromos. Here he became acquainted with the works of Nicomachos and Diophantos. From Diophantos' *Arithmetika,* he learnt those parts which were best understood by his teacher. Reading Diophantos was, after all, never an easy proposition.

The reconquest of Constantinople on the Latins by Michael Palaeologus heralded the start of a Byzantine renaissance. In the thirteenth and fourteenth centuries, intellectual life in the Byzantine Empire intensified and was no longer limited to literature, but also included science and mathematics[55]. During the so-called Palaeologic Renaissance, scholars began to see the inheritance of Greek Antiquity in a different light. The fostering of this inheritance would prove to be of the utmost importance to the fifteenth-century Italian Renaissance.

Two eminent representatives of this Palaeologic Renaissance conducted a first systematic study of Diophantos.

Georges Pachymeros[56] (1242-ca. 1310) first studied in Nicea, but probably ended his work in Constantinople. It is in any case an established fact that he settled in the city after the reconquest. He is best known as the author of a history of the first half century of the Palaeologic dynasty. As a teacher, he wrote a number of treatises, including *Treatise on the four sciences arithmetic, music, geometry and astronomy,* better known as the *Quadrivium.* The chapter on arithmetic is –at least in part– based on Diophantos. The introduction and problems I.1 to I.6 an I.8 to I.11 are all paraphrases, suggesting that Diophantos was used in Byzantine education, even though his influence may have been limited to the rather atypical book I.

The second Byzantine scholar to study Diophantos, Maximos Planudes[57] (ca. 1255-1305), is a key figure in the Greek tradition of the *Arithmetika.* Not unusually during the Palaeologic Renaissance, the well known *literatus* was also versed in sciences and mathematics. Born in Nicomedia in Bythinia (Asia Minor), he moved at a very young age to Constantinople, where he entered a convent. His real name was *Manouel,* but upon entering the convent, he changed it to *Maximos.*

He always took care not to get on the wrong side of the powers that be. On the one hand, he was a supporter of the orthodoxy, as opposed to the Latin cult. On

---

[55]D. GENEAKOPLOS(1984), p.3.

[56]On Pachymeros see N. WILSON(1983), pp.241-242.

[57]C. CONSTANTINIDES(1982), p.62, J. CHRISTIANIDES(2002), pp.155-156.
He argues that, contrary to Tannery's opinion, this does not indicate new methods in the arithmetic education. Until then it seemed that only Nicomachus' *Arithmetica* was being used. Christianides refers to Hierosolymitanus.

the other, he, like cardinal Bessarion, was sufficiently flexible to accept a religious union with Rome whereby the cultural identity was preserved.

He was fluent in Latin, which allowed him to make a (heavily censored) translation of Ovidius' *Metamorphosis* and of texts by Saint Augustine, Boethius and Macrobius[58]. It was also because of his excellent command of Latin that, in 1296, he was sent as an Imperial Envoy to Venice, where he received an honorary citizenship.

More than forty books are attributed to Planudes, covering subjects from theology and grammar to poetry and mathematics. He compiled the astronomical work of Aratus, in which he replaced some of the qualitatively inferior parts with excerpts from the *Almagest*, and he edited Ptolemy's *Geographika*, works by Plutarch, Homer and Hesiod. He was also the author of a book on Indian numbers (Hindu-Arab numerals)[59].

Planudes' interest in the work of Diophantos dates back to 1292-93, as is apparent from his letters from around this period[60]: they tell of how he was collecting Diophantos manuscripts in order to collate them and edit a text that was as complete as possible[61]. At least three Diophantos manuscripts were available to Planudes' circle of friends. One of these was owned by Planudes himself[62], another was in the possession of the astronomer Manuel Bryennios[63]. Planudes asked Bryennios whether he could compare the two manuscripts, which seems to indicate that his own copy was not in the best of conditions. The third manuscript belonged to the imperial library, headed by Theodore Muzalon, who had asked Planudes to restore the copy.

One commentary by Planudes puts the rule of signs in a different perspective (see par. 3.9): "A lacking multiplied by a lacking gives an existing and a lacking with an existing gives a lacking."[64].

As we have previously noted, some authors compare this statement with our sign rule and consequently associate it with the introduction of negative numbers. However, in a commentary, Planudes explains: "He does not simply say 'that what is missing', as if there were a certain presence [and hence something that is missing], but [he speaks of] a presence of which something is missing". In other words, something can only be lacking if something is present: 'that which is simply lacking' does not make sense to Planudes.

---

[58]See W.O. SCHMIDT(1968).

[59]See A. ALLARD(1981b).

[60]P.L.M. LEONE(1991), see letters 66 and 98 among others.

[61]This obviously implies that it was not easy to obtain complete or undamaged manuscripts. If, like Christianides claims, Diophantos was used in education, this should not have been a great problem. His reference to Johannes Hierosolymitanus to assert that Diophantos was used in the eight century would seem to be a backward projection to convince contemporary readers of the quality of the treatise of his protagonist. The fact that these three manuscripts contain just six of the original thirteen books also indicates that the study of Diophantos was limited to say the least.

[62]See A. ALLARD(1981b). This manuscript is probably Ambros. Et 157 sup.

[63]C. CONSTANTINIDES(1982), pp.74, 96 & 142.

[64]Compare with Heath's indication of a literal translation: "A wanting multiplied by a wanting makes a forthcoming." T. HEATH(1964), p.130.

Therefore he can only interpret the rule in relation to what we would call polynomial arithmetic, which makes it possible to deal with expressions such as $(a \pm b)(c \pm d)$.

The two known types of Diophantos manuscripts originated in this period. One of these types contains the commentaries of books I and II and the scholia of the monk Maximos Planudes. These are referred to as the Planudean class of manuscripts[65]. The other class derives from an unknown archetype (see chapter 10). Both versions are apparently drawn from an unknown common archetype.

Four Diophantos manuscripts date back to the thirteenth century, two of which belong to the Planudean class. One of these texts is an autograph of Maximos Planudes[66]. The other Planudean class manuscript was owned by cardinal Bessarion, who donated it to the Marciana library[67]. The two other thirteenth century manuscripts do not derive from a Planudean class manuscript. One was in the collection of the library of the Chapter of the Cathedral of Messina[68]. The provenance of the other manuscript is unknown[69].

Not only did Byzantium preserve a large part of the Greek corpus; this inheritance would also find its way to Europe, as many intellectuals and their book collections travelled west during the fifteenth century. One of these emigrants would build one of Italy's most remarkable libraries.

4.4 Diophantos reinvented: Fibonacci

The first time we encounter Diophantine-like problems in Western European mathematics is in the work of Leonardo of Pisa, better known as Fibonacci (ca. 1170-ca. 1250)[70]. Fibonacci was born in Italy, but grew up in North Africa. His father was a diplomat for the Pisan merchants in Bugia (Bejaia) on the Algerian coast at the mouth of the Wadi Soummam. Fibonacci took mathematics classes, where he was introduced to and learned to work with Hindu-Arab numerals.

He travelled the Arab world until about 1200, when he settled in Pisa. Here he wrote a number of books, including *Liber Abaci* (1202), *Practica geometriae* (1220), *Epistola ad magister Theodorum* (?), *Flos* (1225) and *Liber Quadratorum* (1225). He is also known to have written *Di minor guisa* on commercial algebra

---

[65]C. CONSTANTINIDES(1982), pp.70-73.

[66]A. ALLARD(1979) & (1982-83), p.59. This is Mediolanensis Ambrosianus Et 157 sup.

[67]A. ALLARD (1982-83), p.61. This is Biblioteca Marciana gr. 308., the manuscript only partially dates back to the thirteenth century (see chapter 10).

[68]A. ALLARD(1982-83), p.62. This is Matritensis Bib Nat 4678. Also A. DE ROSALIA(1957-58).

[69]A. ALLARD(1982-83), p.69, this is Vaticanus gr 191.

[70]On Fibonacci and his mathematics, see B. BONCOMPAGNI(1851), E. LUCAS(1877), R. FRANCI & L. TOTI-RIGATELLI(1985), p.18-28, L.E. SIGLER(1987), M. BARTOLOZZI & R. FRANCI(1990), R.B. McCLENON(1994).

and a commentary on Euclid's book X, with a discussion of irrational numbers, but these works have unfortunately been lost.

Given Fibonacci's youth, it should not come as a surprise that these books bear testimony to an Arabic influence. Rashed has demonstrated convincingly that numerous examples in *Liber Abaci* were taken directly from Arab sources[71], while a large part of Abū Kāmil's *On the pentagon and decagon* was used in *Practica geometriae*[72]. Whether Fibonacci was familiar with the work of Diophantos, be it wholly or partially and either in Greek or in Arabic, is unknown. We do however know that he solved problems that were similar to those posed by Diophantos. The manner and order in which he tackles them seem to suggest that he was unfamiliar with the *Arithmetika* - at least in its original form[73], although the possibility remains that he may have encountered originally Diophantine problems in Arab sources. Fibonacci's style, however, has more in common with Euclid's algebraic-geometric approach than with that of Diophantos.

The direct inducement for writing *Liber Quadratorum* came after an invitation to the Imperial Court at Pisa. Fibonacci had come to Emperor Frederick II's attention through his correspondence with scholars at the court, including Michael Scotus (the court astrologer) and Theodorus of Antioch (the court philosopher). During his visit, Johannes of Palermo put him to the test by presenting him with some mathematical problems[74].
One of these problems was: find a solution or an approximation of the solution to the equation $x^3 + 2x^2 + 10x = 20$. Fibonacci had 'demonstrated' (in *Flos*) that the roots of this equations cannot be constructed with straightedge and dividers. It was a first indication that there are other numbers than those defined by the Greeks as a 'constructable number'[75].

---

[71] R. RASHED(2003), pp.55-56.

[72] M. LEVEY(1966), pp.8-9.

[73] The only author who is cited in the *Liber Quadratorum* is Euclid. L.E. SIGLER(1987), pp.13 &96, M. FOLKERTS(2004), p.106.

[74] L.E. SIGLER(1987), p.3, P. VER EEECKE(1952), p.1. According to R. RASHED(2003), p.57, these problems were taken from Arab manuscripts available at the court of Frederick II. Both Theodorus and Johannes were well aware of Arabic mathematics, indeed Theodorus was of Arabic descent and a student of al-Dīn ibn Yūnus. Also R. RASHED(1994b).

[75] The equation can be solved by determining the abscissa of the intersection of the hyperbola $(x + 2)(y + 10) = 40$ and the parabola $y = x^2$.

The solution is $x = \dfrac{\sqrt[3]{352 + 6\sqrt{3930}} + \sqrt[3]{352 - 6\sqrt{3930}} - 2}{3}$.

What Fibonacci did was show that the solution of the equation was neither a whole number, nor a fraction, nor the square root of a fraction. This leads him to the conclusion that it is not possible to solve the problem in one of these manners, and therefore he 'tried to approximate the solution'. This solution is given in the sexagesimal system, which is rather ironic considering that he was the man who introduced decimals in Western Europe. His solution $x = 1.22.7.42.33.4.40 \left(= 1 + \frac{22}{60} + \frac{7}{60^2} + \frac{42}{60^3} + \dots \right)$, which in the decimal system is $x = 1.3688081075$, is correct to nine decimals!

The problem, with the same numerical values, is also encountered in the work of Omar Khayyām. Obviously this equation is of the same kind as those studied by mathematicians of al-Karaji's school[76].

However, it was the following problem that inspired Fibonacci to write *Liber Quadratorum*: *find a square, which when five is added or subtracted is a square again*[77].

In other words $\begin{cases} x^2 + 5 = \alpha^2 \\ x^2 - 5 = \beta^2 \end{cases}$.

Leonardo came up with the correct answer $x = 11\frac{97}{144}$.

The result of his enquiry is one of those shining jewels in the crown of mathematics. Although less well known than *Liber Abaci*, it could be argued that *Liber Quadratorum* is Fibonacci's most intriguing work. It is a well written, beautifully ordered set of theorems and properties about indeterminate quadratic equations, based on elementary relations between square numbers and finite sums of odd numbers. Although it treats similar subjects as Diophantos' *Arithmetika*, it is written in the style of Euclid's number theory works.

The aim of Leonardo is to prove the questions posed by Johannes of Palermo (resolved in proposition 16) and Theodorus (resolved in proposition 20). The book begins with theorems that are reminiscent of the basic properties of the kind of number known to the Pythagoreans, like $n^2 - (n-1)^2 = 2n - 1$, and the very important property $\sum_{i=1}^{n}(2i - 1) = n^2$.

The first theorem[78] is related to Diophantos II.8, but it asks for two squares whose sum is a square, rather than for the division of a number into two squares.

Leonardo begins with a numerical example.
$$\begin{aligned}(1 + 3 + 5 + 7) + 9 &= 1 + 3 + 5 + 7 + 9 \\ 16 + 9 &= 25 \\ 4^2 + 3^2 &= 25\end{aligned}$$

In other words, one should take a sum of successive uneven numbers, the last of which is a square.

Now consider the sum of all uneven numbers smaller than $(2n - 1)^2$, which is an uneven number. The last term in this sum evidently is $(2i - 1)^2 - 2$.

Now
$$\begin{aligned}(2n - 1)^2 - 2 &= 4n^2 - 4n - 1 \\ &= 2 \cdot \left(2(n^2 - n)\right) - 1\end{aligned}$$

Therefore the required sum is
$$\sum_{i=1}^{2(n^2 - n)} (2i - 1)$$

---

[76]M. FOLKERTS(2004), p.105, R. RASHED(2003), pp.57-58.

[77]Compare with al-Khāzin's problem on p. 111. The very same problem was posed in al-Karajī's *Badiʿ*. However, Fibonacci's reasoning differs from al-Karajī's. R. RASHED(2003), pp.68-69.

[78]We use the order as edited by L.E. SIGLER(1987).

Now $$\sum_{i=1}^{2(n^2-n)}(2i-1) = \left(2(n^2-n)\right)^2$$

From which
$$
\begin{aligned}
\sum(2i-1)+(2n-1)^2 &= \left(2(n^2-n)\right)^2+(2n-1)^2\\
&= \left((2(n^2-n))\right)^2+4n^2-4n+1\\
&= \left((2(n^2-n))\right)^2+2.2(n^2-n)+1\\
&= (2n^2-2n+1)^2
\end{aligned}
$$
which is a square.

In theorem 3 Fibonacci proves the well known Pythagorean property
$$\left(\frac{m^2+n^2}{2}\right)^2 = \left(\frac{m^2-n^2}{2}\right)^2+m^2.n^2$$

We then encounter theorems with a distinct Diophantine flavour, such as:
if $a^2+b^2=c^2$, find two numbers x and y for which $x^2+y^2=c^2$ (theorem 5).

Essentially the solution amounts to finding another right triangle with the same hypotenuse as the given right triangle.
Suppose $a^2+b^2=c^2$ and that $m^2+n^2=p^2$ then
$$\left(\frac{m}{p}\right)^2+\left(\frac{n}{p}\right)^2 = 1$$
and
$$\left(\frac{mc}{p}\right)^2+\left(\frac{nc}{p}\right)^2 = c^2$$
Fibonacci distinguishes between three cases, depending on whether the sides are smaller than, longer than or equal to the sides of the given triangle.
The numerical example he gives is $5^2+12^2=13^2$ and $8^2+15^2=17^2$ giving
$$\left(\frac{8}{17}.13\right)^2+\left(\frac{15}{17}.13\right)^2 = 13^2$$
$$\Leftrightarrow \left(\frac{104}{17}\right)^2+\left(\frac{195}{17}\right)^2 = \left(6\tfrac{2}{17}\right)^2+\left(11\tfrac{8}{17}\right)^2$$
$$= 13^2$$

This is followed by the related problem[79] of finding two numbers x and y such that $a^2 + b^2 = x^2 + y^2$.

To solve this problem Fibonacci uses the property he had proven in theorem 6, namely that
$$(a^2 + b^2)(c^2 + d^2) = (ac + bd)^2 + (ad - bc)^2$$
$$= (ad + bc)^2 + (ac - bd)^2$$
Now choose two numbers c and d such that $c^2 + d^2 = k^2$
then write $(a^2 + b^2)(c^2 + d^2)$ as a sum of two squares, e.g. $p^2 + q^2$ leading to
$$a^2 + b^2 = \frac{p^2}{k^2} + \frac{q^2}{k^2}$$
the required numbers are therefore $\frac{p}{k}$ and $\frac{q}{k}$.

These theorems about quadratic sums have, at least in their formulation, a distinctly Diophantine flavour. Other theorems already point toward number theory, for example *if m and n are relatively prime and if m and n are uneven then mn(m + n)(m − n) is divisible by 24, or if m is even and n is uneven then 2m.2n(m + n)(m − n) is divisible by 24*. Fibonacci calls these products congruent numbers and they are important for the rest of his treatise. He shows, for example, that a number of the form $24t^2$ is always a congruent number. A congruent number can however never be a square.

These numbers are important in solving the problem $\begin{cases} y^2 - c = x^2 \\ y^2 + c = z^2 \end{cases}$, which

he interprets as $\begin{cases} x^2 + c = y^2 \\ y^2 + c = z^2 \end{cases}$, in which c is a congruent number.

Fibonacci sees these squares as the sums of uneven numbers and he needs to discuss the number of terms to reach a solution. In the end, after some cumbersome reasoning, he finds the same solution as Diophantos did[80].

In theorem 16, he looks for numbers whose congruent number is a quadratic quintuple. He uses this result to solve Johannes of Palermo's problem in a very general fashion.

---

[79]L.E. SIGLER(1987). This property is known as Lagrange's theorem. Lagrange proved that the product can be written in two, three or four ways as a sum of two squares.

[80]See L.E. SIGLER(1987), pp.53-64 for Fibonacci's text and pp.64-74 for a discussion of Fibonacci's solution. On the history of the problem, see L.E. DICKSON(1971)II, pp. 459 ff.

$$\begin{cases} x^2 + 5 & = & \alpha^2 \\ x^2 - 5 & = & \beta^2 \end{cases}$$

As 5 is not divisible by 24, it cannot be a congruent number and therefore no integer solution exists. Rational solutions can be found by choosing m and n in such a way that their congruent number is a quadratic quintuple.

The following theorems are variants and elaborations of this problem[81]. Other Diophantine-type problems include Theodorus' problem:

$$\begin{cases} x + y + z + x^2 & = & \alpha^2 \\ x + y + z + x^2 + y^2 & = & \beta^2 \\ x + y + z + x^2 + y^2 + z^2 & = & \gamma^2 \end{cases}$$

which he considers as

$$\begin{cases} x + y + z + x^2 & = & \alpha^2 \\ \alpha^2 + y^2 & = & \beta^2 \\ \beta^2 + z^2 & = & \gamma^2 \end{cases}$$

and solves using the theorems for quadratic sums (theorem 20).

One may say without diffidence that Fibonacci was the only European mathematician after Diophantos and before Fermat to make significant contributions to number theory. It was not until Diophantos' manuscripts travelled west to Europe that number theory stood a chance. Slowly but surely, the *Arithmetika* would come to be known in Europe, but not before an event with disastrous consequences for the Byzantine Empire: the fall of Constantinople.

---

[81]In the Florentine manuscript Palatino 577, dating back to the 15th century, variants with other numbers (6, 7, 30) are considered. They are solved using a table of congruent numbers. E. PICUTTI(1979).

Chapter 5

New vistas

As Europe recovered from the Black Death and created a splendid Renaissance for itself, long-held ideas were challenged and replaced with others. The change would manifest itself in literature, the arts and science, though not simultaneously. In this chapter, we consider some of these trends, as insight into the evolution of concepts and worldviews is necessary for an accurate perspective on the story of Diophantos.

5.1   Printed by ...

In the course the 1400s, the printed book gradually gained in popularity, but it was not until the sixteenth century that it came to full prominence[1]. The advent of printing greatly facilitated the dissemination of information. Mathematics, however, remained a field for specialized printers.

Printing was undoubtedly an important factor in the development of Renaissance mathematics and science, though opinion differs on the exact nature of its role.
Producing manuscript copies was a labour intensive process. Inevitably, this meant that texts reached a rather small audience. With the advent of printing, however, it became possible to produce relatively large runs of exact copies at a relatively low cost. Moreover, if demand so required, it was quite easy to produce a second edition. This meant that written materials could be distributed much more widely, which in turn allowed more people to take note of science and its teachings[2].
Books had a further advantage: all copies were identical. Wherever a particular

---

[1] For a discussion of the importance of the book to science and mathematics, see E.L. EISEN-STEIN(1979).

[2] On scientific books and their prices in sixteenth-century Antwerp, see A. MESKENS(1995). The average print run for a more or less scientific book would appear to have been 500 copies, although the number could vary anywhere between 18 and 1200 copies.

copy of a book was read, it contained exactly the same information as all the other available copies. This is not necessarily the case with manuscripts: sometimes scribes would deviate from the original by adding comments or by omitting or editing the text, or by adapting its structure in accordance with their own tastes or preferences. In this sense, every manuscript is unique, unlike a book.

The wider availability of books also freed scholars from the time-consuming chores of copying manuscripts or searching for existing texts on a particular topic.

The undisputed top of the bestseller's list of the era was the Bible, though not necessarily in its Vulgate edition. Protestant Bibles were equally popular. It goes without saying that the authorities were keen to control this new mass medium as best they could. More often than not, publishing a book required licence from the sovereign and sometimes, as in the Low Countries, also from the religious authorities. As the Roman Catholic Church was unable to control printers in Protestant territories, it published an *Index Expurgatorium librorum* or Index of Forbidden Books. This centrally produced *Index* would be supplemented locally. The blacklist could contain some unexpected titles, like books on mathematics for example. This would be due not to the content, but very often to the reputation of the author or the person to whom the book was dedicated. Michael Stifel's *Arithmetica* (1543), for instance, was put on the list of suspect books because it was dedicated to the Protestant pastor of Holtzdorf[3].

Despite their popularity, the Bible and other religious texts did not distract attention from the classical authors. Printers continued to publish technical treatises and, in this way, promoted the dissemination of scientific knowledge. Moreover, competition between publishers stimulated them to try and publish the best books. Some printers actually commissioned scholars to collect and edit material for publication. One such example was Venice-based Aldus Manutius, who established a printing workshop specializing in Greek books. These materials would be edited and prepared for print by Cretan and Greek assistants.

It was an era of a fruitful cooperation between scholars and printers. Indeed, some individuals combined the two careers. Petrus Apianus, a professor of mathematics in Ingolstadt, ran his own printing business, and Oronce Finé, the court mathematician to the French King, also worked as a printer in Paris.

Although practical treatises on arithmetic and astronomy were most prevalent, the classical authors were by no means disregarded. Most editions or translations of classical authors date back to the first half of the sixteenth century. It is in this light that the Diophantos edition by Xylander should be viewed.

Proofreading sometimes brought to light internal inconsistencies. Practical mathematicians (either inspired or compelled by the printing process) had to find

---

[3]A. MESKENS(1994a), p.221.

a common language for expressing their mathematical thoughts, which would constitute the basis for a successful symbolic notation.

Printing was the primary motor behind the (at least partial) standardization of mathematical symbolism, which emerged in the course of the sixteenth century. It also put an end to trade secrets: the centuries-old tradition whereby skills and know-how were transferred exclusively from master to pupil was broken, so that knowledge became a common good.

5.2 Wherefore art thou number?

Slowly, almost unnoticed, a new kind of number came to the forefront of western mathematics: the negative number. It remains an unanswered question whether Diophantos knew negative numbers and regarded them as such. As we have previously mentioned, negative numbers first appeared in Arab mathematics to represent losses, and this is also the form in which they were introduced in European mathematics. Nonetheless, we notice in a tenth-century manuscript that negative numbers as such were already in use. They are indicated in that text by *non existens* or simply by *minus*[4]. The manuscript treats negative numbers as existing quantities, if only to demonstrate how to use the subtrahend. In *Liber Abaci*, Fibonacci considers systems for which the solutions exist only if the parameters are negative. He also discusses these cases and comes to the conclusion that the solutions are positive. If there are negative solutions, he considers the problem as insoluble[5].

The oldest manuscript in which negative numbers are considered acceptable solutions –in a problem where one would expect positive ones– is in a Provencal *Arithmetica* from around 1430.

Nicholas Chuquet appears to have been the first to permit negative solutions in abstract problems (1484)[6]. Luca Pacioli, in the same period, reluctantly accepted negative numbers[7].

Although this reticence towards negative numbers would disappear only slowly, by the sixteenth-century such numbers were widely used in practice. Thus, the way was being paved for yet another kind of number ...

The notion of a "number" has always been enigmatic, particularly in the shift from the countable sets of natural and rational numbers to the uncountable set of real numbers. The question of whether the set of real numbers is a continuum first presented itself –in an embryonic form– in the sixteenth century. We have already discussed the classical notions of number. The query that confronts us now is which transformations does the Diophantine *arithmos* concept have to undergo

---

[4] M. FOLKERTS(1972), p.41.
[5] J. SESIANO(1995), pp.116-133.
[6] J. SESIANO(1995), p.148.
[7] J. SESIANO(1995), pp.134-142.

to become a modern symbolic notion of number? As exploring this matter in any great detail would lead us too far astray from our central topic, we shall consider it only superficially. However, the development of the notion of number has been discussed explicitly by some of the commentators on Diophantos.

Any discussion on the development of number must begin with the Italian *abacists* and their northern counterparts in Germany and the Low Countries[8]. They are the first group of mathematicians for whom we have sufficient material to allow us to form a clear image of their perceptions of mathematics. From the thirteenth up into the sixteenth centuries, we see that their merchants' manuals tend toward greater simplicity in terms of organization and terminology. They concentrate on the practical, everyday uses of mathematics. However far removed from philosophy this may seem, it is here that questions about the nature of number begin to arise. In their work, ratios of numbers are present in the form of solutions to practical merchants' problems, more specifically in problems of company[9]. It was not long before square roots began to appear in merchants' manuals[10]. This leads to problems such as: how does one interpret $6 = \dfrac{6}{3 - \sqrt{3}} + \dfrac{6}{3 + \sqrt{3}}$? The left-hand side obviously is a natural number and therefore a multiple of unity, but what about the right-hand side? It is a sum of two ratios that are incommensurable. To be able to explain this, a different conception was required of the notion of number. It would, however, take another two centuries before a logically sound solution was found, and the first steps toward it would be taken by mathematicians who are part of this arithmetic teaching tradition.

The first indications of a changing conception of number are found in Regiomontanus. He was a man who still stood with one foot in the world of the Ancients, and hence considered numbers as *arithmoi*, or sets of units. However, with the other foot, he stood firmly in the modern world, as all magnitudes are quantities 'that are measured in relation to a certain unit'[11]. According to Regiomontanus, if the surface area of a square is not a perfect square and one wishes to know its side, an approximation is acceptable, 'because it is better to approximate the truth, than to ignore it.'[12].

With Stevin, this reluctance disappears altogether. His definition of number builds on Regiomontanus, but it was also revolutionary, for he dropped the classical definition altogether and accepted the modern notion: "Nombre est cela par

---

[8]On the Italian *abacists* see e.g. W. VAN EGMOND (1976); on the Netherlandish *reken-meesters* see A. MESKENS (1994), (2009) and M. KOOL (1999).

[9]In a problem of company, three or more merchants raise a capital to gain interest or to buy goods. The share of each merchant is different, as is the time during which the merchant is a shareholder and, in some examples, the interest that each receives. M. BARTOLOZZI & R. FRANCI(1990), pp.9-10, A. MESKENS(1994a), pp.66-72.

[10]M. BARTOLOZZI & R. FRANCI(1990), pp.9-10.

[11]A. MALET(2006), p.71.

[12]A. MALET(2006), pp.71-72.

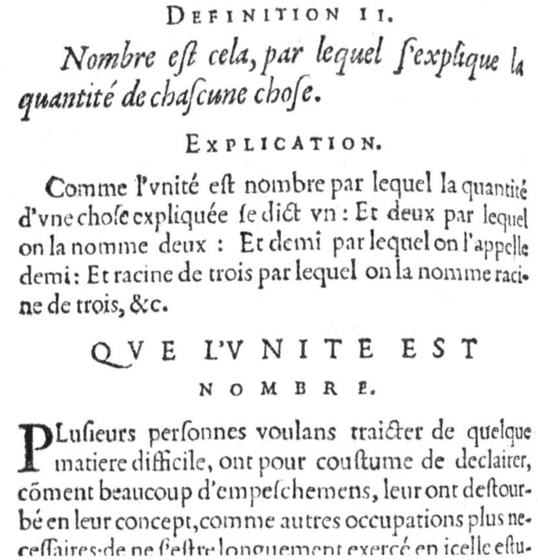

DEFINITION II.

*Nombre eſt cela, par lequel ſexplique la
quantité de chaſcune choſe.*

EXPLICATION.

Comme l'vnité eſt nombre par lequel la quantité
d'vne choſe expliquée ſe dict vn : Et deux par lequel
on la nomme deux : Et demi par lequel on l'appelle
demi : Et racine de trois par lequel on la nomme racine de trois, &c.

QVE L'VNITE EST

NOMBRE.

PLuſieurs perſonnes voulans traicter de quelque
matiere difficile, ont pour couſtume de declairer,
côment beaucoup d'empeſchemens, leur ont deſtour-
bé en leur concept, comme autres occupations plus ne-
ceſſaires de ne ſ'eſtre longuement exercé en icelle eſtu-

Figure 5.1 *Stevin's definition of number. From: Simon Stevin, L'arithmetique. C.
Plantin, 1585. Erfgoedbibliotheek Hendrik Conscience, Antwerpen, G 10413.*

lequel s'explique la quantité de chascune chose", "nombre n'est poinct quantité
discontinue" and "que l'unité est nombre"[13]. Stevin did not consider numbers as a
discontinuous spectrum, but as a continuum. H.J.M. Bos recognizes the influence
of Ramus in Stevin's writings[14]. Ramus (1514-1572) evolved in his thinking about
number. At first he used the classical notion that a number essentially consists of
a number of units. In 1569, however, he wrote in his *Arithmetica* that "number is
that which explains the quantity of each thing."[15], a definition which is borrowed
almost literally by Stevin. Moreover, this definition implies that the unit is a number, which is made explicit by Stevin.

A further explanation lies in the practical mathematical professions that
Stevin came into contact with and where no distinction was made between the
different kinds of number. When constructing a wine gauge, for example, one can
use a rational approximation to indicate a depth point or rely on the construction

---

[13]S. STEVIN(1585), f1 and f.4

[14]H. BOS(2001), p.138. According to R. Hadden, Stevin's notions developed from his dealings
with merchants, especially in relation to the calculation of interest, where the classic conception
is found wanting. The counter argument to this is that numbers that are used in the financial
world are, by definition, rational. An interest is nothing more than a ratio of a hundredth part of
a natural number and another natural (or rational) number. See R. HADDEN(1994), pp.149-154.

[15]"Numerus est secundum quem unumquodque numeratur" (ed. 1569, p.1). See J.J. VER-
DONCK(1966), pp.131-132.

of a square root using Pythagoras' theorem[16].

By interpreting numbers as a continuum, Stevin opened a veritable Pandora's box, the true significance of which he would never fully grasp. He had no idea of the conceptual obstacles that would have to be overcome in order for his insights to be rigorously definable. Yet these insights are crucial in a new interpretation of Diophantos. Whereas Diophantos used only rational –and therefore commensurable– numbers and excluded the incommensurable numbers, including the constructable, now a whole new class of numbers was excluded; numbers whose nature would remain unclear in centuries to come. Nonetheless, it is this new concept of number that distinguishes between algebra and number theory. In a modern interpretation, we may subsume Diophantos with the latter category.

5.3   From the *rule of coss* to algebra

Whereas the Greeks made a distinction between *logistike*, the calculating practices of the merchants and bookkeepers, and *arithmetic*, the science of the properties of numbers, during the Renaissance, the two fields merged. This, together with the rise of printing, was a seminal moment in the history of algebra, lifting the discipline to a higher level as it were. This confluence was preceded by (and partially still coincided with) another development: the use of Hindu-Arab numerals instead of the cumbersome Roman numerals[17].

The fourteenth-century Italian abacists still used rhetoric algebra. Problems and their solutions were given in words, without abbreviations or symbols. The unknown was referred to as *cosa*. The terminology was also used in Southern Germany, where it came to be known as *der Coss*. Sixteenth-century arithmetic books show a diversity in symbolism, an indication that things were on the move. The solution of problems had travelled a long way since the Babylonians. Symbols were attributed to the powers of the unknown. The evolution from rhetoric over syncoptic to symbolic algebra was never straightforward. However, we can roughly distinguish between two schools, namely the Italian and the Southern German school. The symbolism of the latter would come to be used throughout Western Europe.

The first indications of syncoptic algebra are noticeable in Italy. In his *Summa de aritmetica* (1494), Luca Pacioli uses the following abbreviations[18]:

x^0	x^1	x^2	x^3	x^4	x^5	x^6	x^7
n^o	co.	ce.	cu.	ce.ce.	p^o r^o	ce.cu	$2^o r^o$
				censo de	primo	censo de	secundo
numero	cosa	censo	cubo	censo	relato	cubo	relato

---

[16]On wine gauging, see A. MESKENS(1994a), (1994b),(1999).

[17]See for example G.R. EVANS(1977) and W. VAN EGMOND(1976), pp.217-222.

[18]F. CAJORI(1993), pp.107-109.

SIGNIFICATION DES CA-
RACTERES DESQVELS ON
vſe en la Regle d'Algebre.

*Ω.*N.*nombre ſimple ou nombre non denominé*

2. *℞ . radiȝ ou poſition pour la choſe qu'on demande*

4. *℥ .cenſus qui eſt ꝟne ſuperfice quarrée*

8. *℄,cubus qui eſt ꝟn corps de la forme d'ꝟn dé*

16. *℥ ℥ . cens cenſus qui eſt le quadrat d'ꝟn quadrat*

32.*ß.ſur ſoliduȝ qui ꝟient de ℥ en ℄*

Figure 5.2 *Cossic symbols. From: Valentin Mennher,* Arithmétique seconde. *Jan Loë, 1556. Erfgoedbibliotheek Hendrik Conscience, Antwerpen, G 21409.*

This system was used until the first half of the sixteenth century not only throughout Italy by, among others, Tartaglia (*Nova Scientia*, 1537, and other manuals) and Cardano (*Practica arithmeticae generalis*, 1539, 1545, 1570), but also by the Portuguese mathematician Pedro Nuñez[19].

Regiomontanus was one of the progenitors of mathematical symbolism in the west. Already in his manuscripts of 1456, he used symbols for the unknown and its square, for the root, the subtraction and the equality. Whether this symbolism was original is doubtful. It is also used by the monk Fridericus Gerhardt (†1464/65) in a manuscript from 1461[20]. Therefore, one may assume it to have been more or less common in Southern Germany.

x^0	x^1	x^2	x^3	x^4	x^5	x^6	x^7
♄ or N	℞	℥	℄	℥ ℥	ß	℥ ℞	ßß
Dragma or numerus	radix	zensus	cubus	zens de zens	surso- lidum	zensi- cubus	bisur- solidum

[19]F. CAJORI(1993), pp.117-123 and 161-164.

[20]M. FOLKERTS(1977), pp.222-223.

At first sight, the **ϟe** resembles a sloppily written *res*[21], while the **cℯ** seems to be a stenographic abbreviation of *cubus*.

In Western Europe, this notation gained quite a few followers, such as Michael Stifel (*Arithmetica Integra*, 1544, 1545, 1553), Valentijn Mennher and Michiel Coignet (*Livre d'Arithmétique*, 1573), Robert Recorde (*Grounds of the Artes*, 1557), Jacques Peletier (*Algebra*, 1554) and many more[22].

Michael Stifel suggested an alternative system, in which the unknown is represented by a letter. The number of consecutive letters equals the power, thus x^4 would be written as $1AAAA$. The idea of representing the unknown by a letter would, however, not catch on for a few more decades.

Other systems use a symbol that we would refer to as an exponent. Heinrich Schreiber (also known Grammateus) applied such a notation:

x^0	x^1	x^2	x^3	x^4	x^5	x^6	x^7
N	Pri	Se.	3^a	4^a	5^a	6^a	7^a
Numerus	Prima	Secunda	Tercia	Quarta	Quinta	Sexta	Septima

It was also used by, among others, Gielis Vandenhoecke (*In Arithmetica*, 1545)[23]. Again, what we see as an exponent is nothing more than an abbreviation of the name of the power (see fig. 5.3).

In essence, the same system can be found in Rafael Bombelli (*L'algebra*, 1572, 1579)[24], although here a major step has been made in symbolic notation. Bombelli writes ax^n as $\overset{n}{a}$. A similar system is used by Stevin, who writes ax^n as $a(\widehat{n})$. Problematically, Stevin uses the same notation for decimal numbers. For instance $1⓪\ 3①\ 0②\ 2③$ stands for 1.302, while $3① + 2$ means $3x + 2$, which could also be interpreted as $0.3 + 2 (= 2.3)$[25].

Viète uses letters for the unknown, but he continues to name the exponent verbatim or in abbreviation. The names he uses for the powers are the Latin equivalents of Pacioli's names.

Thomas Harriot was the first to use small letters for the unknown (in Stifel's second notation). It is only with James Hume and René Descartes that the system of exponentiation that we use today came into general use[26].

---

[21]Latin for *the thing*. It is the same terminology as Italian *cosa*. See also J. TROPFKE(1980), p.377.

[22]A. MESKENS(1994a), pp.64-66, F. CAJORI(1993), pp.139-147, 164-167, 172-177.

[23]See P. BOCKSTAELE(1985), esp. pp.17 ff.

[24]F. CAJORI(1993), pp.123-128.

[25]On Stevin's symbolism see A. MALET(2008)

[26]F. CAJORI(1993), pp.188-209.

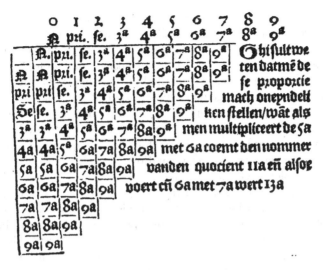

Figure 5.3 *Exponentiation in Vandenhoecke's* In Arithmetica, *Symon Cock, 1545. Erfgoedbibliotheek Hendrik Conscience, Antwerpen, R 50.28*

The polynomial $9x^5 - 7x^4 + 5x^2 - 3x + 1$ thus becomes

with Pacioli	9.p°r°.m.7.ce.ce.p.5ce.m.3.co.p.1.n°
with the cossists	$9\beta - 7\gamma\,\gamma + 5\gamma - 3\mathbf{z} + 1\mathcal{Q}$
with Grammateus	$9\,5^a - 7\,4^a + 5se - 3pri + 1N$
with Stevin	$9⑤ - 7④ + 5② - 3① + 1$
with Viète	A plani cubo 9 - A planoplano 7 + A planum 5 - A3 + 1 N
or	A pl.c.9 - A pl.pl.7 + A pl.5 - A3 + 1N

In these notational systems, equations are solved. They were usually reduced to four known types:

$$ax = b \ (\text{and } ax^n = b)$$
$$ax^2 + bx = c$$
$$ax^2 + c = bx$$
$$ax^2 = bx + c$$

The transformation of equations into cossic or other symbols had an important consequence. By using a notation, even in the semi-symbolic, semi-semantic form of the arithmetic manuals, the attention of the arithmetic teachers was drawn to these imperfections. Although the recreational problems from the arithmetic manuals are founded on a long tradition, they very often rely on implicit assumptions or are undetermined[27].

---

[27] J. HØYRUP(1990b), pp.66-72, cites a problem (p.67) in which the price of a horse has to be determined. It is, however, incomplete and the price can be any multiple of 11. On equations

In this period, a major breakthrough was achieved in the theory of equations. A general algorithm for solving a particular type of third-degree equation was found. Scipioni dal Ferro (1465-1526), a mathematics professor in Bologna, had succeeded in finding a general method for solving equations of the type $x^3 + mx = n$, based on an ingenious substitution of the unknown by two unknowns[28]. As was customary, he kept the method to himself.

In 1535, Niccolo Fontana da Brescia, better known as Tartaglia (the stammerer; 1499-1557), was challenged to solve thirty third-degree equations. Tartaglia, who had claimed that he was able to solve equations of the type $x^3 + mx^2 = n$, succeeded without problems. Girolamo Cardano (1501-1576) persuaded Tartaglia to teach him his method, on condition of total secrecy. However, Cardano was not a man of his word, and published the method in his book *Ars Magna* (1545).

The method of solution led to new problems, opening up a Pandora's box inside a Pandora's box as it were, with square roots of negative numbers[29]. The arithmetic teachers of the time were not aware that they had stumbled upon yet another class of numbers; numbers which, again, would retain their secrets for some decades to come.

---

and algebra in the sixteenth century see also P. FREGUGLIA(2008).

[28]To solve the equation $x^3 + px = q$ $(p, q > 0)$, put $t - u = q$ (1) and $tu = \left(\dfrac{p}{3}\right)^3$ (2)

Equations (1) and (2) lead to a quadratic equation with solutions:

$$t = \sqrt{\left(\frac{q}{2}\right)^2 + \left(\frac{p}{3}\right)^3} + \frac{q}{2} \text{ and } u = \sqrt{\left(\frac{q}{2}\right)^2 + \left(\frac{p}{3}\right)^3} - \frac{q}{2}$$

The solution of the equation becomes: $x = \sqrt[3]{t} - \sqrt[3]{u}$, which is easily checked by substitution.

[29]Consider the equation $x^3 = 15x + 4$, put $t + u = 4$ and $tu = \left(\dfrac{15}{3}\right)^3 = 5^3$

This leads to the quadratic equation $t^2 - 4t + 125 = 0$ with a negative discriminant. The solution would then be $x = \sqrt[3]{2 + \sqrt{-121}} + \sqrt[3]{2 - \sqrt{-121}}$. This is a *casus irreducibilis*. We know there is a solution: $x = 4$ (the other solutions are $x = -2 - \sqrt{3}$ and $x = -2 + \sqrt{3}$). One can prove that there are *apparently* no real solutions if, in the equation $x^3 = px + q$, $27q^2 - 4p^3 < 0$, when in fact there are three real roots! If $27q^2 - 4p^3 > 0$ then the equation has one real and two complex roots. In other words, with this method one finds the real root if complex roots are present, but one cannot find any of the real roots if there are three.

Chapter 6

Humanism

6.1   Trait d'union: Bessarion and the humanists

Early in the fifteenth century, long before 1453, the westward exodus of scholars, from the Byzantine Empire to Italy, had already started. This was to give a
decisive impetus to the renewed interest in Classical Antiquity, which had been
manifesting itself in Italy since the fourteenth century. It led to the spectacular
growth of humanist libraries[1]. In a mediæval library, one would typically find only
Euclid and Archimedes, usually in their Latin translations, whereas in a humanist
library, one would encounter not only Euclid and Archimedes in both Latin and
Greek, but also numerous mathematical texts by other authors.
After the establishment of a chair of Greek in Florence in 1396, the town grew to
become the centre of the study of Greek codices. The most important catalyst for
the study of Greek in Italy was the Council of Florence in 1438-39. The purpose of
this Council was the unification of the Roman and Byzantine Churches. The *fine
fleur* of Italian humanism attended: Bruni, Traversari, Poggio, Valla and Nicholas
of Cusa[2].

Concerted efforts were made to locate further Greek codices. Fra Ambrogio
Traversari, for example, made a tour of the Italian abbeys, while Poggio Bracci-
olino's search focused on the trans alpine abbeys[3].
Giovanni Aurispo (1376-1459) and Francesco Filelfo (1398-1481) travelled to Byzantium to report on the state of its libraries and they brought home with them numerous manuscripts. Aurispo, after his second voyage, is believed to have brought
back no fewer than 238 codices, and Filelfo apparently took home some fifty

---

[1]See P. KIBRE(1946) and P.L. ROSE(1973).
[2]J. GILL(1961), pp. 85 ff.
[3]On these book hunts, see for example P.W.G. GORDAN (1974). On Poggio, see E.
WALSER(1974).

A. Meskens, *Travelling Mathematics - The Fate of Diophantos' Arithmetic*, Science Networks.
Historical Studies 41, DOI 10.1007/978-3-0346-0643-1_6, © Springer Basel AG 2010

manuscripts[4]. With these new acquisitions, the Italian libraries soon became the
most important in the Western world.

It was during the reign of Nicholas V (1397-1455, pontiff from 1447) that the Roman Renaissance began. Nicholas had set himself the task of building the world's
largest library, equalled only by the ancient library of Alexandria. Moreover, the
fall of Constantinople in 1453 prompted an influx of Greek manuscripts. By 1455,
the Vatican library contained 414 Greek manuscripts, twelve of which were of a
mathematical or astronomical nature. Among these texts is 'a medium-sized volume on papyrus, bound in red leather and entitled *Arithmetika* by Diophantos of
Alexandria'[5] and the *Tables* of Theon of Alexandria[6].

Sixtus IV (1414-1484 - pontiff from 1471) continued Nicholas's library policy and
soon two further versions of Diophantos also found their way to the *Vaticana*[7].

Equally important to the Italian humanist libraries was the private manuscript collection of Cardinal Basileios Bessarion, the embodiment of the merger between East and West[8]. Born into a family of manual labourers in Trebizond on the
Anatolian Black Sea, his intelligence caught the attention of the local metropolitan, who arranged for him to be sent to Constantinople to study. Bessarion would
travel the Peloponnese, explore Neoplatonism under Gemistos Pletho and devote
himself to a broad range of mathematical fields. In Byzantium, he advocated much
needed reforms, inter al. to the future emperor Constantine XII. In view of the
(technical) superiority of the Latin culture, he argued that Constantinople should
attract metallurgists, mechanics, arms manufacturers and shipbuilders from Italy.
His suggestions were however not well received and would never be put into practice.

In 1437, he was appointed Metropolitan of Nicaea. The following year he
travelled to Italy, to attend the Council of Florence. His countrymen resented his
notion that the Catholic and Orthodox Churches could be reconciled. Pope Eu-
gene IV, on the other hand, appreciated his viewpoints and invested him with the
rank of Cardinal. From 1440, he lived in Italy. It was Bessarion who, in a decisive
way, made mathematics an integral part of the *studia humanitatis*[9]. He had an
immense, possibly even the largest, private collection of Greek manuscripts and
mathematical texts, which in 1468 he donated to Venice, where it would become
the nucleus of the Biblioteca Marciana. Among these manuscripts were Greek
codices by Apollonios, Archimedes, Aristarchos, Diophantos and seven books by
Euclid, including not only the *Elements* but also the *Optics* and *Catoptrics*[10].

---

[4]P.L. ROSE(1976), p.28.

[5]Vat.Gr. 304. P.L. ROSE(1976), p.37.

[6]E. MÜNTZ & P. FABRE(1887), p.339, P.L. ROSE(1976), p.37, R. McLEOD(2000), p.9.

[7]Vat. Gr. 191 and 200, P.L. ROSE(1976), p.38.

[8]The seminal study on Bessarion is the three-volume treatise by L. MOHLER(1923).
Other studies on Bessarion and fellow-emigrants can be found in J. MONFASANI(1995).

[9]D. GENEAKOPLOS(1984), p.23 and P.L. ROSE(1976), pp.44-46, 49, 56.

[10]P.L. ROSE(1976), p.45, L. LABOWSKI(1979), no. 118 in inventory 1474, p.198.

With the exception of Pappos, his collection contained all classical mathematical writings.

In collaboration with Cardinal Bessarion (1403-1472), Nicholas V established an Academy for the study of Greek philology. The aim of the Academy was the translation into Latin of Greek and Byzantine books. Bessarion was driven by a certain patriotism and the hope that it would contribute to mustering support for a new crusade[11]. Hitherto, the Italians had given preference to corrupted Greek texts and second or third-hand Arab translations over the more or less immaculate Byzantine originals (among other things because, since the Schism, the Byzantines had been regarded as heretics)[12].
The Academy consisted of Greeks and members of the Curia alike. Some actually travelled to Constantinople to learn Greek[13].

Bessarion, aided by the *Academy* and the patronage of scholars, succeeded in promoting the translation of all major scientific Greek treatises into Latin.

6.2   Diophantos goes north: Regiomontanus

Through his contacts with Cardinal Bessarion, Regiomontanus' interest in Greek mathematicians, and Archimedes in particular, grew. Regiomontanus (1436-1476) was born as Johann Müller in Koenigsberg (present-day Kaliningrad). Aged eleven, he enrolled at the university of Leipzig, where he would study astronomy and mathematics. Later, in 1450, he went to the university of Vienna, attracted by the reputation of Georg Peurbach (1423-1461)[14].

In May 1460, Cardinal Bessarion arrived in Vienna on a papal mission to gain imperial support for the war against the Turks[15]. Ironically, it was in this period that the last Byzantine bastion against the Turks, his hometown of Trebizond, fell (1461). By this time, Bessarion had begun to translate Ptolemy's *Almagest*. He hoped that he could improve on what he felt was an inferior translation by George of Trebizond[16]. He invited Peurbach and his student Regiomontanus to travel back

---

[11]C.L. STINGER(1985), p.120.

[12]D. GENEAKOPLOS(1984), p.61.

[13]D. GENEAKOPLOS(1984), p.17.

[14]M.H. SHANK(1996), M. FOLKERTS(1996). On Regiomontanus' mathematical work in Vienna, see M. FOLKERTS(1980) and (1985).

[15]M. FOLKERTS(2002), p.44.

[16]George of Trebizond (Trabzon) was one of the most important translators of the *Academy*. At the insistence of cardinal Bessarion, he had added a commentary to Ptolemy's. This commentary was based on, but sometimes also contradicted, Theon of Alexandria's commentary. Although there was a twelfth-century version of the *Almagest* by Gerard of Cremona, George of Trebizond's version would become the standard (see D. GENEAKOPLOS(1984), p.19.). The fact that Bessarion and George of Trebizond were at odds over the topic of Platonism undoubtedly clouded Bessarion's opinion. On this controversy, see for example J. MONFASANI(1976),

𝕭effarion cardial' grec°

𝕵ohannes de monte re gio aftronimus

Figure 6.1 *Cardinal Bessarion (l) and Regiomontanus (r). From: Hartmann Schedel, Das Buch der Chroniken und Geschichten. Anton Koberger, 1493. Erfgoed-bibliotheek Hendrik Conscience, Antwerpen, B 481.*

to Italy with him, hoping that, with their assistance, he would be able to finalize the translation. Peurbach died before Bessarion's departure, but Regiomontanus was easily persuaded and joined the Cardinal's retinue for the following years[17]. Regiomontanus finished Peurbach's *Epitomi Almagesti* and wrote *De triangulis omni modis libri quinque* (Five Books on All Kinds of Triangles). By this time, Bessarion had taught him fluent Greek.

Some time before 1467, Regiomontanus travelled to Hungary, where he joined the court of King Matthias Corvinus. Here, he drew up trigonometrical tables and solved astronomical problems. From 1471 to 1475, he lived in Nuremberg, where he also opened a print shop. In 1476, he travelled to Rome to take part in discussions on calender reform. It would be his last journey, as he died unexpectedly in Rome that same year.

The real interest in Diophantos in the West only began in earnest after Regiomontanus had visited the Venetian library and had found six of Diophantos' books in its collections. He wanted to translate them, but only in conjunction with the other seven. He therefore asked one of his Italian correspondents to enquire with the *Graecisti* in Ferrara about the other volumes, but without success[18].

passim.

[17]On Regiomontanus' Italian years, see R. METT(1989).

[18]J.-A. MORSE(1981), p.62.

Probably in 1464, he gave a series of lectures at Padua University, of which only the introductory *Oratio* has come down to us. In this *Oratio*, he presents his views on the history of mathematics, with emphasis the continuity of the mathematical tradition, as is apparent from the transfer of knowledge from culture to culture and from language to language. He puts it that mathematics began with the Egyptians, who had to redivide the banks of the Nile after the annual flooding. This geometrical practice culminated, after having been transferred to the Greeks, in the work of Euclid. For arithmetic, he goes as far back as Pythagoras and then draws a line from Euclid and Jordanus of Nemore to the Renaissance. Shortly before his 1464 lectures, he had encountered the Diophantos manuscripts, which suggested to him that the origin of algebra lay in Antiquity. Thus, Diophantos came to epitomize the transmission of knowledge over the centuries and between cultures[19], although the emphasis on the humanist ideal of the rebirth of classical knowledge may have prejudiced Regiomontanus in forming his opinion. According to him, the work of Diophantos contains the "whole flower of arithmetic, the *ars rei et census*, which we now refer to under its Arab name of algebra"[20].

In his letters to Giovanni Bianchini, Regiomontanus poses problems which have an unmistakable Diophantine flavour to them. It is doubtful, though, whether this is a direct consequence of the discovered manuscript, as he had, in 1456, already posed similar questions[21]. Two of the problems he presented to Bianchini were:

> find four squares whose sum is again a square[22] and find three squares that are in harmonic proportion.

The further correspondence does not reveal whether Regiomontanus himself was able to solve the problems, though it seems self-evident he was. It was, for that matter, customary for mathematicians and arithmetic teachers to present as a challenge to their peers problems to which they themselves already knew the answer[23].

---

[19]P.L. ROSE(1976), p.96 and J.-A. MORSE(1981).

[20]F. SCHMEIDLER(1972), p.46, G. CIFOLETTI(1992), pp.260-261, J.S. BYRNE(2006), p.55.

[21]M. FOLKERTS(2002), p.414, (1996), pp.105-108 and (1985), pp.207-219. This manuscript (Plimpton 188) is largely an autograph by Regiomontanus, in which we find Johannes de Muris's *Quadripartitum numerorum*, Gerard of Cremona's translation of al-Khwārizmī's *Algebra* and sixty-four problems presumably written in 1456. Among these, we encounter indeterminate systems of equations and systems with more Diophantine-like problems, such as
$$\begin{cases} x+y+z &= 214 \\ x^2 - y^2 &= y^2 - z^2 \end{cases} \text{ and } \begin{cases} x+y+z &= 116 \\ x^2 + y^2 + z^2 &= 4624 \end{cases}$$
and remainder problems.

[22]This problem is solved by Diophantos by using II.8. If
$$\begin{cases} x^2 + y^2 &= \beta^2 \\ z^2 + w^2 &= \gamma^2 \\ \beta^2 + \gamma^2 &= \alpha^2 \end{cases}$$
is true then
$x^2 + y^2 + z^2 + w^2 = \alpha^2$ is also true. The system is easily solved by applying II.8 in which $\gamma = k\beta - \alpha$, $x = ky - \beta$ and $z = lw - \gamma$.

[23]See for example A. MESKENS(1994a), pp.81-84.

Regiomontanus also opened a print shop in Nuremberg[24]. He had the intention of publishing all mathematical and astronomical treatises that seemed important to him. To this end, he drew up a publication plan, which he also published and which mentions not only Diophantos' *Arithmetika*, but also Jordanus of Nemore's *Arithmetica* and *Numeris Datis* and Johannes de Muris's *Quadripartitum Numerorum*, as well as the anonymous *Algorithmus demonstratus*[25]. His untimely death prevented him from fully implementing his plan.

Decades later, Regiomontanus' attempt to publish an edition of Diophantos came to the attention of Wilhelm Holtzman (1532-1576), better known as Xylander, who would become the first to attempt to publish *and* edit the book.

---

[24] A. WINGER-TRENNHAUS(1991).
[25] M. FOLKERTS(2002), p.413.

Chapter 7

Renaissance or the rebirth of Diophantos

7.1 Xylander: A sphinx to solve a riddle

Regiomontanus' discovery remained hidden for about half a century. There are but few indications that Diophantos was studied during the first half of the sixteenth century. Yet it was the sixteenth-century interest in ancient texts that led to his proverbial rebirth. It was, after all, a period when study groups on classical authors were established. One such group centred around Paolo Manuzio (1512-1574), the son of the printer Aldus Manutius, who specialized in Greek editions. In his circle, we find Gian-Vincenzo Pinelli, a bibliophile who owned at least one Diophantos manuscript, Nicaise van Ellebode[1] (ca. 1535-1577), a Flemish student at Padua University, and Andras Dudith (1533-1589), a Hungarian scholar who resided in Italy for the Council of Trent[2]. Dudith left Italy in 1562, after having been appointed as Bishop of Csanàd. He later became Bishop of Pecs and Sziget, two cities that, like Csanàd, were located in Turkish-occupied territory. In 1565, he was sent by Maximilian II as an imperial envoy to Krakow. During this period, he maintained a lively correspondence with Manuzio's study group. Two years later, however, it emerged that he had married, which shocked the Italians to the extent

---

[1] Nicaise van Ellebode, or Helbault, came from a poor family in Kassel. He studied at Louvain University from 1549. In Cardinal Granvelle he found his first patron. In 1552, he began to study at the Collegium Germanicum in Rome. He subsequently became a teacher at the Academy of Tyrnau (Hungary), where he met Andras Dudith. During the 1560s, he returned to Italy to study philosophy and medicine, and later also Greek. In 1577, at the request of Bishop Radéczi, he moved to Pressburg, where he soon died of the plague. He made several translations of Greek authors, including Aristotle, Aristophanes and Apollonios, but only one was published (*De natura hominis liber unus*, Plantin, Antwerp, 1565; MPM A398, EHC D2286). On van Ellebode, see D. WAGNER(1973).

[2] J.-A. MORSE(1981), p.192; on Manuzio, see also P.L. ROSE(1976), pp.190, 194, 196, 198, 218.

that they ceased all correspondence with him. In 1568, he was excommunicated[3]. Previously in Italy, Dudith had managed to locate many of Pinelli's manuscripts –some of which he copied– but not the *Arithmetika*. Around 1570, however, he wanted to obtain a copy of the work. Apparently, his interest had been aroused by reading Regiomontanus (most probably the 1464 letter to Bianchini). As he was unable to contact Pinelli directly, he asked Nicaise van Ellebode to act as an intermediary[4]. Van Ellebode forwarded the question to Camillo Zanetti in Venice[5], who complied by sendings him a copy. Dudith was convinced he had received a copy of the Pinelli manuscript, as is confirmed in van Ellebode's correspondence. However, the copy cannot be traced to any existing Pinelli manuscript[6], so that the Dudith version was, probably via an unknown copy, derived from a manuscript by Matteo Macigno, which Camillo Zanetti had compiled in 1560-1565. This manuscript is based on two progenitors: a fifteenth-century non-Planudean and a fourteenth-century Planudean manuscript[7].

Wilhelm Holtzman (1532-1576), or Xylander in Greek, was a professor of Greek and logic at the University of Heidelberg. He translated many works, including Dio Cassius, Plutarchos and Strabo into Latin. He also translated the first six of Euclid's books into German and Michael Psellos' *Quadrivium* into Latin.

In October 1571, he paid a visit to some friends at the University of Wittenberg, a call that would not be without consequences. During a conversation with the mathematics professors Sebastian Theodorich and Wolfgang Schuler, the topic turned to Diophantos, of whom the latter possessed a corrupted text they were unable to correct[8]. On his journey home, Xylander found a solution to one of the problems the text posed. As he passed through Leipzig, he discussed his method with Simon Simonio. Simonio and Joannes Praetorius subsequently acted as intermediaries to ensure that Xylander could borrow the Dudith manuscript in order to study it in Heidelberg[9]. In less than three years, Xylander succeeded in translating the book from Greek into Latin and in publishing the text. The work not only contains the translation, but also scholia and commentaries by Xylander. Each type of text has its own type face, which makes it easy to distinguish between the units. The problems are set in bold Roman type; Planudes' commentaries carry the heading SCHOLION and are set in small italics, while his own commentaries are marked XYLANDER and set in large italics. Wherever Planudes had missed a mathematical point, Xylander corrected the former's errors and also digressed upon the points made by Diophantos. In most cases, however, Xylander felt he

---

[3]A. ALLARD(1985), pp. 299, 312.

[4]P. COSTIL(1935), p.296. Letter of Nicaise van Ellebode to Adriaan van der Myle.

[5]On Zanetti as a calligrapher and copyist, see R. CESSI(1925).

[6]See A. ALLARD(1985), with an extensive description of the manuscript and its antecedents.

[7]A. ALLARD(1982-83), pp.76, 82-83. The manuscript is based on *a Mediolanensis Ambrosianus A91 sup* and *T Vaticanus gr. 304*.

[8]W. XYLANDER(1575), introduction.

[9]W. XYLANDER(1575), introduction, A. ALLARD(1985), pp.297-298 and 309-310.

needed to supplement Planudes' notes, sometimes even on a different topic[10].
In his translation, Xylander freely applies the symbolic notation in use in his region at that time. However, he also explains the procedures used in his own words.
Let us again take problem II.8 as an example.

II. *Divide a square into two squares.*

Diophantos puts that if $x^2 + y^2 = 16$,
then also $y^2 = 16 - x^2$ and puts $y = 2x - 4$.

$$
\begin{array}{r}
2N - 4 \\
2N - 4 \\
\hline
- 8N + 16 \\
4Q \quad - 8N \\
\hline
4Q \quad - 16N + 16
\end{array}
$$

This has to equal $16 - Q$ or $5Q - 16N \parallel 16$,
from which $5Q + 16 \parallel 16 + 16N$,
This immediately implies the equality of $5Q$ and $16N$ and therefore of $5N$ and 16.

In passing, Xylander also notes that $(2n + 1)^2 - (2n)^2 = 2n + 1$.

It goes without saying that Xylander's version would not pass the scrutiny of contemporary text editions. The syncoptic notation of Diophantos is translated, without any ado, into symbolism, thus $\bar{\mu}\bar{\beta}\varsigma \uparrow \delta^v \bar{\alpha}$ becomes $16 - Q$ and $\varsigma \bar{\beta} \uparrow \bar{\delta}$ becomes $2N - 4$.
If the Greek text was incomprehensible, he made no attempt to clarify the text, but simply rendered it in equally incomprehensible Latin. It was only in his commentaries that he solved the unclarities, sometimes with the relevant Greek texts alongside. For Xylander, an edition was nothing more than the start of a series of commentaries on the mathematical and philosophical contents. In this sense, he was an exponent of the old school. However, this is not to minimalize Xylander's work. He stood at the cradle of the emergence of critical text edition, and his work has undoubtedly helped to make Diophantos' work accessible to a broader audience. Moreover, the Diophantos manuscripts that were available to him were corrupted: the numbers were often incorrect and the train of reasoning was hard to follow. In Xylander's own words, he had to be a sphinx to solve the riddles associated with the manuscript[11].

---

[10]See also J.-A. MORSE(1981), p.199.
[11]W. XYLANDER(1575), p.94, comment added to III.16 (III.15 in Tannery's numbering). The deplorable state of the manuscripts is confirmed by Allard's research (1980)(1985), who found no fewer than 300000 errors in thirty-one manuscripts of a text containing only about 58000 words, which averages out at roughly one error per five words.

Only two years after Xylander's publication, Guillaume Gosselin used it for his own *De Arte Magna [...] quae & Algebra et Almucabala, libri quatuor*[12]. Almost simultaneously, Diophantos was rediscovered in Italy, not by a philologist, but by a mathematician. He would give the text another, more mathematical, interpretation putting literary considerations aside.

7.2   Coincidence of traditions: Rafael Bombelli

Around the time when Xylander was preparing his text edition of Diophantos, a second Diophantos manuscript came to the attention of another mathematician. He would open up a new avenue in the history of Diophantine analysis and make the elusive author known to a broader audience of scholars.

About a century earlier, Regiomontanus had argued that algebra had classical roots. In his day, however, algebra was a practical art, pursued by arithmetic teachers and engineers, not by the university-educated audience he was addressing. Whereas mathematics scholars had no problem bridging the gap between the two worlds, most practical mathematicians did, because more often than not they neither read nor understood Latin. Hence, Regiomontanus' oratio rather predictably did not meet with a wide response.

The man who would change this was himself a practical mathematicians. His name was Rafael Bombelli (1526-1572/73).

Few biographical details are known about Bombelli[13]. His father was called Antonio Mazzoli, but he changed his surname to Bombelli. Antonio, a wool tradesman, was married to Diamante Scudiere. The couple had six children, the eldest of whom was Rafael. Rafael got an education with the engineer and architect Pier Francesco Clementi. He found a patron in Alessandro Rufini, the later bishop of Melfi.

In 1549, Rufini obtained the privilege to drain the swamps of the Val di Chiani, in the Papal States. Bombelli worked on this project from 1551 to 1555, when it was temporarily suspended. To fill this period of inactivity, he set himself the task of writing a comprehensive yet accessible algebra book. The material he intended to use was not new. In some cases, it had already been published or was considered common knowledge. In his opinion, however, previous books were inadequately arranged, and he thought it should be possible to unify the various techniques, rules and constructions that were applied. He wanted the text to be self-contained and accessible even to those who had had no mathematical education beyond elementary arithmetic. Bombelli began writing his treatise in 1557. By 1560, the work in Val di Chiani had been concluded, and Bombelli left for Rome, where he was consulted on the draining of the swamps in Lazio, a persistent source of malaria since Roman times. As Bombelli recounts, it was during this project that 'a Greek manuscript, compiled by Diophantos, was found in the Vatican library.

---

[12]SBA G4825, MPM a692.2(2) and P. VER EECKE(1926), pp.LXXI-LXXV.
[13]J.-A. JAYAWARDENE(1963) and (1965); J.E. HOFFMAN(1972), pp.196-197.

It was shown to me by Antonio Maria Pazzi, public professor of mathematics in Rome. We have begun the translation and have already finished five of the seven existing books. We have not been able to finish the others because of other commitments.'[14]. Shortly after the publication of his *Algebra*[15], which contains these translations, Bombelli died (1572/73).

Algebra consists of three books, the first two of which deal with algebra while the third is concerned with number theory. Two further books, dealing with geometry, were prepared but never published. *Algebra* is dedicated to Alessandro Rufini, the Bishop of Melfi and Bombelli's patron. In his dedication, Bombelli describes algebra as higher arithmetic, invented in India and introduced in Europe via Arabia. Remarkably, he also mentions Diophantos, who does not fit into this picture. This may be due to the fact that he only learnt of the existence of Diophantos after he had written the first draft of his manuscript.

The reference to Diophantos gives Bombelli a strong argument for a reappraisal of algebra within the field of mathematics. As in the work of Regiomontanus a century earlier, it was also an invitation for the Humanists to devote some attention to the book: by stressing the Greek roots of algebra, he detaches the discipline from the tradition of craftsmen and arithmetic teachers and brings it into the realm of humanist scholarship. Yet Bombelli's experience as an engineer is apparent throughout the work, because he stresses the aspect of problem-solving methods rather than the underlying mathematical structures or techniques. Be that as it may, Bombelli brought Diophantos within reach of all mathematicians, without breaking with the algebraic tradition.

A comparison between the manuscript (most probably dating from 1557-1560)[16] and the published version of the book shows that the discovery of the Diophantos manuscript influenced Bombelli's thinking. Books 1 and 2 would remain virtually unchanged, but whereas the manuscript version of book 3 consists entirely in typical arithmetic-teaching problems, this is by no means the case for the printed version. In the first book, Bombelli introduces terminology and symbolism. Although he attributes his symbolism to Diophantos, he does not deviate from the Italian conventions for powers. Diophantos' $\kappa^v \delta$ becomes *primo relato* (the first posed) and $\kappa^v \kappa$ becomes *the cube of the power*. Bombelli went up to the twelfth power in this fashion. However, his interpretation is multiplicative, unlike Diophantos' additive notation. Whereas the Italian arithmetic teachers used words like *cosa* for the unknowns and *censo* for the power, Bombelli uses terms which derive from the Diophantine terminology: *tanto* and *potenza*.

He also discusses the operations with numbers, including square roots. It is important to note that Bombelli appeared to have no objections to using $\sqrt{-1}$ and that

---

[14]P.L. ROSE(1976), pp.146-147; R. BOMBELLI(1963) , *Algebra* sig d 2r-v.

[15]R. BOMBELLI(1963).

[16]There are two known manuscripts: one complete (in the library of Bologna) and one containing books 3 and 4 (university library of Bologna). See S.A. JAYAWARDENE(1973).

he even provided rules for manipulating this kind of number. They are used in the second book to solve equations. As a curiosity, they are also used as imaginary *tanto* to quadratic equations. In solving cubic equations, they are of course encountered when dealing with 'irreducible equations'. In present-day terminology, he calculates the cubic root of conjugate complex numbers by accepting that they are again conjugate[17].

The practical problems of book 3 have disappeared and have been replaced with 270 other problems, 143 of which were selected from the *Arithmetika*. The others are similar to those originally devised by Bombelli[18]. The numbers proposed by Diophantos have more often than not been changed, so we cannot speak of a translation as such. The initial, linear, problems are intended to clarify the restatement of a verbal problem, as a calculation that has been stripped entirely of the terminology typically applied by arithmetic teachers.

Bombelli's treatment of systems of quadratic equations begins with Diophantine problems, but he also refers frequently to methods proposed by Pacioli and Cardano. Again, he adds his own problems, in which he does not limit himself to rational numbers, but also considers irrational square and cubic roots. The indeterminate problems have been selected from the first five books of the *Arithmetika*. Not only have the proposed numbers been changed, but so too has the method of solving the problems. Bombelli often adds the general rule, in the traditional terminology of the arithmetic teacher. These are usually word-by-word descriptions of the algorithms that need to be followed, with references neither to the unknown nor to the numerical values. In this manner, Bombelli is able to formulate the Diophantine problems more generally. An advantage of this method is that it can also be applied to algebraic numbers, that is to say numbers with an unknown quantity.

To find two numbers in a given proportion and with a given sum, one proceeds as follows[19]:

> Add the two numbers of the proportion and divide the given number [= the sum of the nominator and denominator in a proportion of a fraction equal to the given proportion] by the sum. Multiply the quotient by both numbers of the proportion and these two products are the sought after numbers.

Bombelli then gives an example with numbers whose proportion is 2 to 3 and whose sum is $x + 5$. By applying the rule, it becomes clear that $x + 5$ must be divided by $2 + 3 = 5$. The numbers, then, are resp. $\frac{2}{5}(x+5) = 2\left(\frac{x}{5}+1\right)$ and

---

[17]I. BASHMAKOVA & G. SMIRNOVA(2000), pp.73-75.

[18]For a concordance between the problems in Diophantos and Bombelli, see K. REICH(1968).

[19]R.BOMBELLI (1963), p.321. This procedure amounts to: $\begin{cases} \dfrac{x}{y} = \dfrac{m}{n} \\ x+y = p \end{cases}$.

From which $x = \dfrac{m}{n}y \Rightarrow \dfrac{m}{n}y + y = p \Rightarrow y = \dfrac{np}{m+n}$.

$$\frac{3}{5}(x+5) = 3\left(\frac{x}{5}+1\right).$$

Whereas Diophantos considers indeterminate equations in preparation of further problems, Bombelli regards them as problems in their own right.
In problem V.16, Bombelli demonstrates he has complete command of the Diophantine methods. This problem asks for three numbers the sum of whose cube minus any one of those numbers is again a cube.
After Diophantos has found the equations, he must divide 162 into a sum of three cubes. Diophantos notices that $162 = 125 + 37 = 125 + 64 - 27 = 5^3 + 4^3 - 3^3$. The Diophantos manuscripts state rather enigmatically that 'we find in the Porisms that the difference of two arbitrary cubes can be rendered as [...] cubes'. Here, Bombelli explicitly transforms $37 = 4^3 - 3^3$ into a sum of two cubes. As he only uses one symbol for the unknown, his procedure is sometimes confusing[20].

Suppose $37 = 4^3 - 3^3 \quad = (4-t)^3 + (t-3)^3$
Now $\qquad\qquad (4-t)^3 \quad = 64 - 48t + 12t^2 - t^3$
and $\qquad\qquad (t-3)^3 \quad = -27 + 27t - 9t^2 + t^3$

Replace t by $\dfrac{27}{48}t = \dfrac{9}{16}t.$

The sum $\left(4 - \dfrac{9}{16}t\right)^3 + (t-3)^3$ which has to equal 37.

This equation is simple to solve, as the linear terms cancel each other out.

And the final result is: $37 = 4^3 - 3^3 = \left(\dfrac{40}{91}\right)^3 + \left(\dfrac{303}{91}\right)^3.$

Bombelli, therefore, did not just translate Diophantos; he also adapted, edited and amended the problems, which clearly indicated that he was a match for Diophantos. In a sense, Bombelli did nothing that commentators before him had not done, yet his approach was boldly new. He refused to be a slave of the text, as Xylander was. His approach to the *Arithmetika* is that of a mathematician. If he needs to clarify, he does so by adding a properly chosen example, in which the problems faced are solved. He also puts the problems in a wider mathematical perspective, points out that a typical Diophantine problem has no unique solution, and gives a general algorithm to solve the problem. It would therefore appear that Bombelli was the first Renaissance mathematician to fully comprehend and appreciate the Diophantine corpus. Thanks to Bombelli's treatment, Diophantine

---

[20] See J.E. HOFFMAN(1972), pp. 213-214. The general problem can be solved as follows, using two unknowns. Put $a^3 - b^3 = (a - x)^3 + (y - b)^3$ and put $a^2x = b^2y$ (1), then the equation becomes $y^3 - x^3 = 3(by^2 - ax^2)$ (2). From (1) it follows that we can put $x = \dfrac{b^2}{t}$ and $y = \dfrac{a^2}{t}$. Substituting in (2), we find $\dfrac{1}{t} = \dfrac{3ab}{a^3 + b^3}$, from which the solution immediately follows. For the actual problem see: E. BORTOLOTTI(1963), pp.453-454.

analysis entered a new phase: it was finally placed in the mathematical footlight again, after slumbering in near-darkness for almost a millennium and a half. Unfortunately, Bombelli's work did not reach as wide an audience as it deserved. Sometimes, however, quality is more important than quantity, for there was at least one scholar who read both Xylander and Bombelli. His name was Simon Stevin, and he was one of the finest mathematicians of his era.

7.3   The great art: Guillaume Gosselin

During the sixteenth century, numerous arithmetic books were published across Western Europe. As it was cited in Johann Scheubel's *Algebra* (1550), Regiomontanus' view on the history of mathematics and algebra, which afforded a prominent place to Diophantos, became widely accepted [21]. Diophantos' manuscripts also became known in France. Petrus Ramus refers to them in *Schola mathematica* (1569), again with Regiomontanus' views as a starting point. Ramus' *Schola mathematica* tried to radically alter the history of mathematics by suggesting that there was a direct lineage leading from Ancient Greece to Western Europe. According to him, transmission occurred via Byzantium and Italy[22].

The first mathematician to use Xylander's work was, most probably, Guillaume Gosselin. Although not well known, Gosselin is a key figure in the development of theoretical algebra, which would culminate in the work of Viète. Biographical data on Gosselin is scarce. We know that he was born in Caen, but not in which year. He is referred to as 'young' in one of the dedicatory poems of his edition of Tartaglia's *Arithmetica* from 1578. This leads Cifoletti to conclude that he was not yet thirty at the time, implying that he was born after 1548[23]. He began to work at the court in Paris at a very young age, probably after being introduced by a relative called Jean Gosselin. Jean Gosselin was a librarian with the Royal Library and he also served as court mathematician to Margaret de Valois, Queen of Navarre[24].

During his time in Paris, Guillaume stayed at the Collège de Cambrai. He became involved with the Académie de Baïf, a group centred around Jean Antoine de Baïf (1532-1589) that studied music and mathematics[25].
Within this academy and in other intellectual circles, there was a notion – not

---

[21]G. CIFOLETTI(1992), p.126.

[22]J. HØYRUP(1996), pp.114-115, also R. GOULDING (2006).

[23]G. CIFOLETTI(1992), p.54.

[24]Sister of Henry III and wife of Henry IV.

[25]The *Académie de Poésie et de Musique* was the first French academy to be established by royal decree. In 1570, Charles IX, Jean Antoine de Baïf and Joachim Thibault de Courville (c. 1530-1581) asked permission to found an academy for the purpose of reviving Graeco-Roman poetry and music. The academy also had a moralistic undertone, inspired by Neoplatonism. The link with mathematics lies in Pythagorean philosophy, which puts it that the physical universe can be rendered as numbers and posits a harmony between the universe and the structure of the

unlike, mutatis mutandis, that in Italian humanist milieus a century earlier – that France should take part in the Greek, classical Roman and Italian cultures. The first reference to Gosselin as a mathematician is in an *Oratio* (1576, but delivered in 1575) by Maurice Bressieu (ca. 1546-1617)[26]. His name is mentioned alongside several other acquaintances of Bressieu whom the author finds significant to mathematics in France. Bressieu is believed also to have worked on an edition of Heron's writings, on the basis of copies at the Royal Library[27].

Gosselin wanted to publish an edition of Diophantos' work. He is known to have published three books with the house of Gilles Beys (1542-1595)[28], but none on Diophantos. For the preparation of such an edition, Gosselin seems to have had access to the otherwise rather closed Royal Library, which possessed a Diophantos manuscript[29]. Perhaps this was due to the fact that the edition was to be part of a larger project, conducted under the patronage of influential individuals such as Renaud de Beaune and Auguste de Thou (1553-1617)[30], aimed at making the ancient mathematical corpus more easily accessible to contemporary readers. In his book *De Ratione*, Gosselin reveals details of his edition. Apparently, he was entrusted with the editing task by François Viète, Jacques Cujas (1520-1590)[31] and Jacob Holler[32]. Through the offices of Cardinal Jacques Davy du Perron (1556-1618)[33], Gosselin obtained a copy of the Vatican Diophantos manuscript. Unfortunately, Gosselin's manuscript has not been preserved[34].

From the very first pages of his book *De Arte Magna* (1577), Gosselin refers to Diophantos and uses some of his problems. Gosselin's symbolism is closely related to that of Pacioli, but it is typographically simpler. For powers he uses the multiplicative form, fully aware that he diverts from Diophantine usage[35]. The contents of the first two books is classic for a sixteenth-century algebra book. It deals withroot calculating, proportions, the rule of three, problems relating to the

---

human soul. F. YATES(1947), pp.21 and 38, also G. CIFOLETTI(1992), p.57.

[26] Bressieu became a mathematician at the Collège de France in 1575. In 1586, he was appointed as the King's representative to the Holy See. Here, he became steward of the Vatican Library.

[27] Letter from Gosselin to Bishop Renaud de Beaune (1527-1606), who was *maître des requêtes* of the Parisian parliament and a very influential figure at the court.

[28] Son-in-law of the Antwerp printer Christopher Plantin and head of the Paris branch of the latter's firm.

[29] Probably Parisinus gr. 2380.

[30] A historian and politician, and a member of an influential French family. He collected material with a view to the compilation of a history of France.

[31] A French lawyer and professor with different institutes. He was *conseiller au parlement de Grenoble*.

[32] A lawyer and parliamentarian.

[33] Confidant of Charles IX, Henry II and Henry IV. He was a Calvinist, but converted to Catholicism in 1577. He took his vows in 1593 and was appointed to the position of Cardinal and Archbishop of Sens. He was a member of the Council of Regents (1610).

[34] G. BACHET(1621), p.4 *Ad Lectorem*.

[35] G. GOSSELIN(1577), p.4.

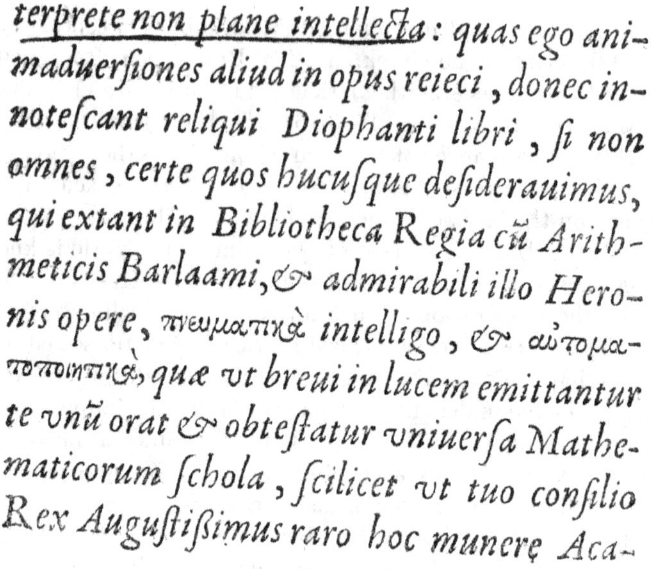

terprete non plane intellecta : quas ego ani-
maduerfiones aliud in opus reieci , donec in-
notefcant reliqui Diophanti libri , fi non
omnes , certe quos hucufque defiderauimus,
qui extant in Bibliotheca Regia cū Arith-
meticis Barlaami, & admirabili illo Hero-
nis opere, πνευματικὰ *intelligo , &* αὐτομα-
τοποιητικὰ, *quæ vt breui in lucem emittantur*
te vnū orat & obteftatur vniuerfa Mathe-
maticorum fchola , fcilicet vt tuo confilio
Rex Auguftißimus raro hoc munere Aca-

Figure 7.1 *From the introduction of Guillaume Gosselin, De Arte Magna, Gilles Beys, 1577. Erfgoedbibliotheek Hendrik Conscience, Antwerpen, G 4825.*

regula falsi[36] and the double *regula falsi*[37], and calculating with polynomials[38]. The third book is devoted to equations ordered by degree and in second order by the Diophantine method of solution. After a discussion of equations of the third degree, he describes selected equations from the *Arithmetika*, referred to as the *fictitia aequatio*. These are indeterminate equations in which an expression is to be equalled with a square or a cube. His first example is "$6x^2 + 16 = $ a certain square". His solution is, of course, based on Diophantine methods.

Choose a square for which the root is a binomial in x, e.g. $2x + 4$.
The square of which has to equal $6x^2 + 16$
We thus find $6x^2 + 16 = 4x^2 + 16x + 16$, from which $x = 8$.

Gosselin also uses the binomial $3x - 4$ to demonstrate the possibilities and the limitations of the method.

---

[36]Using the *regula falsi*, the root of an equation is found by substituting two arbitrary numbers. If x_1 gives a difference of f_1 and x_2 a difference f_2 then the solution is $x = \dfrac{x_1 f_2 - x_2 f_1}{f_2 - f_1}$.

[37]The double *regula falsi* used to solve two simultaneous equations in two unknowns. See J. TROPFKE(1980), pp.371ff.

[38]H. BOSMANS(1906), pp.47-55

A second Diophantine method is that of the *duplicata aequatio* or double equation (see p. 83), which is used for finding the three terms of an arithmetical sequence whose terms are squares[39].

> If the difference is 96, then the terms are $x^2, x^2 + 96$ and $x^2 + 192$
>
> The difference between the last two is of course 96.
>
> Now $96 = 4.24 = 6.16 = 8.12 = p.q.$
>
> The solution is given by $\left(\dfrac{p+q}{2}\right)^2 = x^2 + 192$ and $\left(\dfrac{p-q}{2}\right)^2 = x^2 + 96$.
>
> With $p = 8$ and $q = 12$, we find a negative x^2, which is excluded.
>
> With $p = 4$ and $q = 24$, we find $\left(\dfrac{1}{2}.28\right)^2 = 196 = x^2 + 192$,
>
> therefore $x^2 = 4$ and $x = 2$.

Altogether new is that some Diophantine problems –from the atypical first book– are solved using two unknowns.

> *Divide 100 into two parts such that a fourth part of the first number exceeds a sixth part of the second part by 20* (Diophantos I.6).

> Call the numbers $1A$ and $1B$ then $1A + 1B = 100$ and $\dfrac{1}{4}A = \dfrac{1}{6}B + 20$.
>
> Therefore $1A = \dfrac{4}{6}B + 80$ and, considering that $1A + 1B = 100$, we
>
> can equal $1A$ to $\dfrac{4}{6}B + 80$, so that $\dfrac{5}{3}B + 80$ equals 100. If we omit the
>
> unnecessary, we find $\dfrac{5}{3}B = 20$. We divide 20 by $\dfrac{5}{3}$, yielding 12, which
>
> is the number B, so that the number A equals 88.

The transition to two unknowns (and implicitly a system of simultaneous equations) is a deviation from and – in terms of legibility – an improvement on the original Diophantine solution.

Although Gosselin indicated on more than one occasion that he was preparing an edition of Diophantos, the project never materialized. Could it be that another edition interfered with this project? After Xylander's translation by Bombelli, Diophantos would soon be published in a contemporary mathematical form. Moreover, Gilles Beys's father-in-law, Antwerp-based printer Christopher Plantin, had published an own version of Diophantos. Perhaps he dissuaded Gilles Beys to market a rival French edition. And with this, we have arrived in the Low Countries.

---

[39]See 3.7, p. 83. See also H. BOSMANS(1906), pp.60-61 and G. CIFOLETTI(1992), pp.129-130.

7.4   The marvel is no marvel: Simon Stevin

The first edition to be conceived not so much as a text-critical study, but rather as a mathematical translation with respect for the original problems was the edition by Simon Stevin. Stevin was, without any doubt, one of the best practical mathematicians of his age. Born in Bruges in 1548, he was the illegitimate child of Anthuenis Stevin and Cathelyne van der Poort. Cathelyne later married Joost Sayon, a member of a wealthy Bruges mercantile family that traded with, among other places, the Baltic Region. Stevin worked first at the tax office of the Brugse Vrije[40] and subsequently as a bookkeeper and cashier in Antwerp. His life is relatively easily reconstructed from 1581 onwards. In 1581, he was admitted as a freeman of Leyden. Here, he developed a friendship with Prince Maurice, who at that time was studying at Leyden University. From 1584, Stevin would act as a *praeceptor* to Prince Maurice for his geometrical studies. Later, he was appointed as a counsellor. He is the author of several books, some of which were published posthumously by his son[41].

A first productive period in his authorship was during the 1580s. His first book, *Tafelen van interest* (Interest tables), was published by Plantin in 1582 in Antwerp. This was followed the next year by *Problematum Geometricorum*, published by Jan Bellerus, also in Antwerp. The religious strife that raged through the Low Countries prompted Plantin to move his print workshop to Leyden, and it was there that Plantin published the book that would make Stevin's name: *De Thiende*, which would later appear in a French edition entitled *La Disme*[42]. In that same year, Plantin also printed *Dialecticke ofte bewysconst* (a book on logic) *and L'Arithmétique* [43].

L'Arithmétique is a contemporary and competent compilation of sixteenth-century algebraic knowledge. Much of the material was already known and is comparable to that found in other arithmetic books of the time. However, the book by Stevin does have a certain originality, as the author explains his new symbolism for writing unknowns in equations (see par. 5.3) and allows negative coefficients in equations[44]. He discusses the equations $x^2 = ax + b$, $x^2 = ax - b$ and $x^2 = -ax + b$. While these equations were also dealt with in other arithmetic books, the latter only allowed positive coefficients, i.e. $x^2 = ax + b$, $x^2 + b = ax$ and $x^2 + ax = b$.

Appended to *L'Arithmétique*, we find Stevin's translation of Diophantos, which took as its starting point Xylander's Latin version[45] and is comparable in

---

[40]The hinterland of Bruges.

[41]More details about Stevin's life can be found in E.J. DIJKSTERHUIS(1970) and G. VANDEN BERGHE(2004). On the mathematics in the work of Stevin, see A. MESKENS(1996) and H.J.M. BOS(2004).

[42]*The Tenth*, a book on the use of decimal notation.

[43]See D. IMHOF(2004).

[44]H.J.M. BOS(2004).

[45]One may wonder whether Stevin ever came into contact with Andras Dudith prior to his publication of Diophantos. In a now untraceable letter from Ortelius to Justus Lipsius, we read:

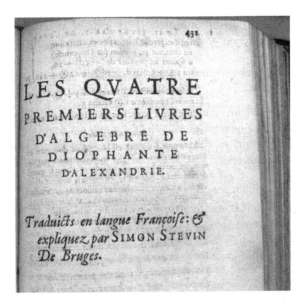

Figure 7.2 *Stevin's Diophantos. From:* L'Arithmetique, *C. Plantin, 1585. Erfgoedbibliotheek Hendrik Conscience, Antwerpen, G 10413.*

approach to Bombelli's treatment. Contrary to Bombelli and Gosselin, however, he does not restrict himself to a selection of problems, but provides a full mathematical translation of the first four books of Diophantos' *Arithmetika*. The last two books were added in a later edition of 1625 by Albert Girard. Stevin, in his own words, explains that he was prevented from translating them by more urgent business.

Stevin was, for that matter, not interested in producing an accurate translation, which he felt was, in any case, impossible due to the corruption of the texts. He was more preoccupied with adapting the problems and solutions to the style encountered in the rest of *L' Arithmètique*[46].

Stevin thus incorporated Diophantos entirely into the algebraic corpus and moved him away from number theory. He was aware that the problems in books II to IV are formulated in a general way and often have an infinite number of solutions. Instead of allowing this, he introduced the numbers that Diophantos merely uses

---

'he has told Dudith what he wrote to Plantin about Simon, the mathematician, but Dudith has not yet replied.' Dudith had asked Justus Lipsius, the famous Antwerp humanist, to act as an intermediary to persuade Stevin to come to Breslau. Lipsius was very sceptical about the 'mathematician' (whose name is replaced by asterisks in his published letters), "He is a mere mathematician [mathematicus erim merus] without any other craftsmanship, indeed, any knowledge of languages, in short the type one considers more as an applied scientist than as a theoretician." Clearly, then, Lipsius is anything but complimentary about Stevin! On Stevin and Lipsius' letter to Dudith, see R. DE SMET(2004).

[46]S. STEVIN(1625), pp.407-408.

as an example in the problem statement. This way, he succeeds in reducing the general problem to a particular problem, which of course limits its scope.

If we take problems II.8 and II.9 as an example, we notice how he still demands that the numbers should be rational (*commensurables*)[47].

8. *Partons une nombre quarré à sa racine commensurable, comme 16, en deux semblables quarrez.*
Divide a square number that is commensurable with its root, like 16, into two similar squares.

Let the first number be 1②, which makes the second $-1$② $+16$.
This is a square that is commensurable with its root whose side we equal to a number of times ① $- \sqrt{16}$, for example 2① $- 4$ and the square is 4② $- 16$① $+16$.
When reduced, this gives 5① equal to 6 and according to the sixty-seventh problem 1① equals $\dfrac{16}{5}$.
I say that $\dfrac{256}{25}$ and $\dfrac{144}{25}$ are the required squares.
[...]
Note: It is known that one can find an infinite number of right triangles whose sides are an arithmetical number. If one is asked for the side that contains the right angle, the root of the said $\dfrac{256}{25}$ is equal to $\dfrac{16}{5}$ and the other side the root of $\dfrac{144}{25}$, which is $\dfrac{12}{5}$, and the hypotenuse will be 4.

9. *This ninth question is the same as the eight, but shall be solved in a different way.*

Let the side of the first square be 1① and the side of the second square a number of 1① $- \sqrt{16}$, for instance 2① $- 4$.
So the first square is 1② and the second square is 4② $- 16$① $+16$.
The sum of the squares therefore is 5② $- 16$① $+16$ and, considering the sixty-seventh problem, 1① will equal $\dfrac{16}{5}$.
So $\dfrac{256}{25}$ and $\dfrac{144}{25}$ are the two required squares, as above.

---

[47]What Stevin means by a similar square is a square commensurable with its root. II.9 is Diophantos' second solution to the problem. Thus, Diophantos II.9 becomes Stevin's II.10.

In the tradition of arithmetic books, Stevin then gives the general rule, e.g. for a double equation

Rule

One finds two numbers whose product equals the difference of two given numbers, but on the condition that the square of half their difference is larger than the smallest given arithmetical number, or, which amounts to the same, the square of the half of their sum is larger than the largest given arithmetical number.

Problems that are solved by Diophantos by means of the double equation (see p. 83) are solved in similar fashion by Stevin, e.g. problem II.12

Find a number which, when 2 and 3 are added, becomes a square.

Suppose the number is 1② − 2, then (1② − 2) + 2 is a square. If we add 3 to this number, we find 1② + 1. This number has to be a square. Equal this to the square of 1① − 4. Then 1② + 1 equals 1② + 8① + 16 and ① equals $\frac{15}{8}$, which solves the problem.

In these first four books, we find one problem that has been literally translated, and it is not altogether clear why. Was it perhaps overlooked in the manuscript and subsequently appended under time pressure, perhaps by someone else than Stevin?
Stevin, on the other hand, succeeded in making a very strong case for his notation. Although inferior to the notation later introduced by Descartes, it is nonetheless clearer than the cossic notation. Moreover, this Diophantos edition is the first that, while faithful insofar as mathematical content is concerned, translates the first four books into contemporary mathematical terms. To Stevin, the mathematical content takes precedence over literary criticism. This fresh approach to Diophantos, unlike the earlier literary edition, was conducive to new mathematical insights.

Girard's translation (1625) of books V and VI is inferior to Stevin's, even though he had access to Bachet's masterly rendering and edition of Diophantos, as is apparent from a number of additions by Girard. For example, he sometimes asks for solutions in the integers instead of the rationals. And he does not always adhere to Stevin's notational system, thereby breaking up the unity of the text so carefully crafted by the latter. Girard uses some of Viète's solutions in the latter's notation. Viète, for his part, had his own particular insights into the meaning of the *Arithmetika*.

Chapter 8

Fair stood the wind for France

8.1 Diophantos' triangles: François Viète and the *New Algebra*

François Viète was born in Fontenay-le-Comte in 1540. After an education at the Franciscan school of Fontenay, he enrolled at the university of Poitiers in 1558 to study civil and canonical law. It took him just one year to obtain a baccalaureate and a licence degree, after which he embarked on a successful career as a lawyer in his hometown. In 1564, he became a secretary to Jean and Antoinette de Partenay, a position he combined with tutoring their daughter Catherine (1554-1631) in mathematics. Catherine would later marry Charles de Quellenec. After a quarrel with her son-in-law, Antoinette moved to La Rochelle, a stronghold of the Huguenot movement, and Viète followed her. It was here that he came into contact with figures from highly influential Huguenot circles, including Henry of Navarre, the later King Henry IV, and his niece Françoise de Rohan, to whom he became a legal counsellor.

In 1571, Viète went to Paris and became *avocat au parlement*. Here, he witnessed the St. Bartholomew's Day massacre, during which Charles de Quellenec, the husband of his former disciple, was killed. Viète's pupil Catherine owed her life to the actions of her brother René de Rohan. Viète himself seems never to have gotten in any danger: apparently he succeeded in adopting a neutral position during this time of religious strife[1].
In 1573, he was appointed by Charles IX as a member of the parliament of Brittany. He moved to Reims, where he would live for the next seven years. In March 1580, Henry II called him back to Paris to act as a counsellor. However, religious tensions were still running high in the French capital, including at the royal court.

---

[1] According to K. REICH & H. GERICKE(1973) Viète remained Catholic, J.J. O'CONNOR & E.F. ROBERTSON (s.d.) maintain that he was a Huguenot.

Viète fell under suspicion of being a Huguenot, and his presence was no longer desired[2].

Viète retreated to Beauvoir-sur-Mer. During his five-year exile there, he had enough spare time on his hands to devote himself to his favourite pastime: mathematics. It is in this period that his most important mathematical ideas began to take shape.

After a turbulent period in 1587-88[3], Viète became a counsellor and cryptographer in the service of Henry IV. As a cryptographer, he was able to decipher an intercepted letter from the Spanish King and French Pretender Philip III to the Archduchess Isabella[4], which helped Henry thwart Philip's military plans.

In 1597, Viète went on special leave in his hometown of Fontenay-le-Comte only to return to his position in Paris barely two years later. In 1602, an exhausted Viète left the service of Henry IV. He died a year later.

Despite the fact that, as an attorney, he was merely an 'amateur', Viète was one of the great mathematicians and he carried out important work in, among other fields, the theory of equations. It should be noted that Viète 'homogenized' his equations. This means that he reduced all terms of an equation to the same 'kind'. He would have regarded the equations $x^3 + x = 8$ as $x^3 + 1^2x = 2^3$. Thus, each term in the equation represents a rectangular parallelepiped. In the original equation, a cube is added to a line to find a line, which he considered absurd. He distinguished between numbers as numbers and numbers as geometrical entities. If two numbers are multiplied, the result is a number; if, however, two lines are multiplied, the result is a rectangle (area).

Viète introduced a method which, by means of an appropriate substitution, can be used to solve second, third and fourth-degree equations (we ignore the homogenization here).One may wonder to what extent Viète was inspired by Diophantos' substitutions (see par. 3.7)

---

[2]Due mostly to machinations of the so-called Holy League, which schemed to keep protestants as far away from power as possible.

[3]In 1584, Henry of Navarre became the legal heir to the throne, in what would prove to be the start of a bitter struggle for power. The ensuing war saw several political assassinations. The principal victims were Henry III and Henry, Duke of Guise. Henry IV eventually ascended the throne in 1589. He immediately had to contend with a strong Spanish intervention. Not until 1593, when he 'returned to the Roman Catholic faith', was he able to enter into Paris. He is said to have declared *'Paris vaut bien une messe'*. Viète supposedly reconverted around this time, but it is doubtful whether he was ever a Huguenot in the first place.

[4]The then governor of the Spanish Netherlands.

For the equation
$$x^2 + bx = c$$

make the substitution
$$y = x + \frac{b}{2} \text{ or } x = y - \frac{b}{2}$$
then

$$y^2 = x^2 + bx + \left(\frac{b}{2}\right)^2 \Leftrightarrow y^2 = c + \left(\frac{b}{2}\right)^2$$

which immediately gives the value of x.

For the third degree equation $x^3 + bx^2 + cx + d = 0$,

he puts
$$x = y - \frac{b}{3}$$
yielding the equation
$$y^3 + py + q = 0$$

A second substitution
$$y = z - \frac{p}{3z}$$

which has become the classical substitution for solving third-degree equations, yields a biquadratic equation in z^3

$$z^3 - \frac{p^3}{27z^3} + q = 0$$

from which
$$z^3 = -\frac{q}{2} \pm \sqrt{\left(\frac{p}{3}\right)^3 + \left(\frac{q}{2}\right)^3}.$$

Viète only uses the positive cube root of z, but it is easily demonstrated that the six solutions for z yield three different solutions for x.

For the fourth-degree equation, Viète uses the substitution
$$x = y - \frac{b}{4}$$
which reduces the general equation to

$$x^4 + px^2 + qx + r = 0$$

which can be solved using Ferrari's method.

It goes without saying that finding these substitutions encouraged Viète and other mathematicians to search for similar substitutions for higher-degree equations, an undertaking that was, for that matter, doomed to failure.

Inspired by Diophantos' *Arithmetika*,Viète suggested that perhaps algebra could be used to solve geometrical and arithmetical problems[5]. He intended to

---

[5]See J.-A. MORSE(1981), W. VAN EGMOND(1985), P. FREGUGLIA(1989) & (2005) and H.J.M. BOS(2001).

explain his reconstruction of the classical solution methods in a series of books entitled *New Algebra*.

His approach consisted in three parts: *zetetike, poristike* and *exegetike* (or *rhetike*), the three consecutive steps in algebraic problem-solving.
Zetetike is the art of translating the problem from a geometrical or algebraic formulation into an equation in one or more unknowns, as in the *Arithmetika. Poristike* are techniques for transposing these equations or ratios into other equations or ratios. These transformations may result in conditions of existence (cf. diorismos) or in another, more general, problem. Examples of this can be found in Theon of Alexandria and Archimedes. *Exegetike* is the way in which algebraic or geometrical solutions are found from the equations drawn up with *zetetike* and transformed with *poristike*. Viète was, however, unable to provide a classical example of this.

In 1591, he began to work on a project intended to result in a series of books that together would constitute the *Opus restitutiae mathematicae analyseos seu algebra nova (Book of the restored analysis or new algebra)*[6]. The series was to consist of ten books, but only seven were ever published.
In one of these books, entitled *Zeteticorum libri quinque*, Viète turns his attention to Diophantine problems to demonstrate his new solution method. The book is undated and it is often encountered in a single volume alongside *In Artem Analyticem Isagoge* (1591). Research by Warren van Egmond suggests it was printed in two parts[7]. The first eight sections were probably printed by Jamet Mettayer in Tours, as the paper on which it appears is similar to that used in other books by this printer. The other sections were printed on a different kind of paper: the same as was used for *De numerosa potestatum resolutione*, which was printed in 1600 by David Leclerc.
As we have previously noted, *zetetike* is the translation of a problem into an equation. By referring to Diophantos, Viète, much like Regiomontanus and Bombelli before him, was able to give the topic of his work an aura of respectability and tradition, reaching back to Antiquity. Yet Viète was of the opinion that Diophantos did not apply *zetetike* correctly. Diophantos, he argued, failed to solve the general problem, but instead used specific cases to illustrate his solution procedures. Consequently, the procedure itself tends to be obscured, which would not have been the case had he worked with kinds rather than specific numbers. Viète proposes to clarify this in *Zeteticorum*. Algebra had undergone an evolution since the time of Diophantos, so that the latter's text had become unrecognizable. Algebra had matured, necessitating a reinterpretation of the *Arithmetika*. About a third of *Zeteticorum* was borrowed from Diophantos' *Arithmetika*[8].

---

[6]W. VAN EGMOND(1985), pp.367-368.
[7]W. VAN EGMOND(1985), p.362.
[8]See K. REICH & H. GERICKE(1973).

Viète is not the only reader or commentator to have claimed that Diophantos's style is rather obscure. This assessment is in part due to the fact that we are unable to read the *Arithmetika* through the eyes of a mathematical contemporary of Diophantos. The criticism usually relates to three aspects. Although Diophantos promises a general solution, he solves the problem using specific values. If Diophantos encounters difficulties in using these values, he, without distinction to the solution of the actual question, first solves an easier problem needed to solve the original query. For the unsuspecting reader, Diophantos' approach can be somewhat enigmatic. The proposed numbers and parameters do indeed lead to a solution, but the reasoning behind their selection is never explained. The reader can only guess as to whether or not Diophantos had a general algorithm at his disposal.

In Viète's algebraic application, whereby the geometrical approach often confounds the actual algebra, Diophantos' problems are resolved in a general way. His structure of proof is such that there are no specific problems to resolve on the way, just ordinary equations. Viète does pay a price, though, because what he does is beyond the realm of Diophantine number theory.

Viète provides a geometrical interpretation to arithmetical problems, often using rectangular triangles. In *Notae Priores* (1631), he indicates that two numbers (A, B) can produce a Pythagorean triplet $(A^2 + B^2, 2AB \,|A^2 - B^2|)$. He also shows that, if (Z, B, D) and (X, F, G) are Pythagorean triplets, then
$$(XZ, |FB \pm DG|, |BG \mp DF|)$$
is also a Pythagorean triplet[9]. He frequently uses these properties when solving Diophantine problems.

Let us consider Viète's interpretation of Diophantos II.8 (in Viète's *Zeteticorum IV.1*[10]). Viète gives two solutions: the second refers to the Diophantine solution, while the first is Fibonacci's. In the latter, he makes use of a number triangle, i.e. a right-angled triangle with two known sides.

---

[9]If $a^2 = k^2 + l^2$ and $b^2 = m^2 + n^2$ then
$$
\begin{aligned}
a^2.b^2 &= (k^2 + l^2)(m^2 + n^2) \\
&= k^2m^2 + k^2n^2 + l^2m^2 + l^2n^2 \\
&= k^2m^2 + l^2n^2 \pm 2klmn k^2n^2 + l^2m^2 \mp 2klmn \\
&= (km \pm ln)(kn \mp lm)
\end{aligned}
$$
A classical proof for this proposition is given by Bachet in Porism II.7. Diophantos most probably also uses the proposition in III.65 in which he states that $65^2 = (3^2 + 2^2)(2^2 + 1) = 4^2 + 7^2$.
I. BASHMAKOVA & I. SLAVUTIN(1976/77) interpret Viète's construction as a predecessor to the multiplication of complex numbers. Although a parallel does exist, their explanation is not credible. The right triangles may be considered as complex numbers $z = b + di$ and $x = f + gi$. The product xz then produces one of Viète's triangles.
Note that even Fibonacci used a method that is essentially the same as that described here.
[10]F. VIÈTE(1646), p.62.

Suppose the square of F has to be divided into two squares.

Choose a right number triangle with hypotenuse Z, base B and perpendicular D.
The solution is given by constructing the similar right triangle with hypotenuse F.
Because Z is to F as B is to the base and D is to the perpendicular, the base is $\dfrac{BF}{Z}$ and the perpendicular $\dfrac{DF}{Z}$
The sum of the squares of these numbers is the square of F.

If we follow the Diophantine reasoning to divide B^2 into two squares, then we put one side equal to A and the second to $B - \dfrac{S}{R}A$.
The sum of the squares, then, is:
$$A^2 + B^2 - \frac{2SAD}{R} + \frac{S^2A^2}{R^2}$$
which has to equal B^2, from which
$$A^2 - \frac{2SAD}{R} + \frac{S^2A^2}{R^2} = 0$$
and $A = \dfrac{2SRB}{S^2 + R^2}$ and as second side
$$C = \frac{\left(R^2 - S^2\right)B}{R^2 + S^2}$$

With the numbers R and S, we construct the number triangle
$$\left(S^2 + R^2, 2SR, R^2 - S^2\right).$$
So B is to $S^2 + R^2$ as A is to $2SR$ and C to $R^2 - S^2$.

To divide 100 into two squares, we make a right number triangle with 4 and 3, which makes the hypotenuse 25, the base 7 and the perpendicular 24. Then 25 is to 7 as 100 is to 28 and 25 to 24 as 100 to 96. Making the square of 100 equal to the square of 28 plus the square of 96.

Viète always starts from a Diophantine problem, which he proceeds to solve in a general way, using his own methods and techniques. The homogenization of equations sometimes forces him to rephrase the problems. It is however clear that Viète interprets Diophantos in a highly original, geometric fashion.

8.2 Emulating the Ancients: Claude-Gaspar Bachet de Méziriac

Xylander's translation of the *Arithmetika* was conceived neither as a definitive nor even a correct interpretation. Moreover, his edition of the Greek text was never published. His addition of commentaries suggests he adhered to the old belief that, in text editing, the addition of consecutive commentaries can clarify the intention of the author.

Two generations later, this viewpoint had become obsolete. Consecutive commentaries were omitted in favour of a readable, definitive text in Greek as well as in (Latin) translation. Texts were studied from a philological, historical and mathematical point of view, and unclarities, errors and corruptions were resolved. In this respect, the text of *Arithmetika* offers the advantage that it is quite stereotypical, so that corrections depend primarily on the mathematical insight of the translator or editor rather than on their philological prowess.

A first attempt at editing the Greek text was undertaken by Joseph Auria, about whom we know very little. He is believed to have lived in Naples around 1590, where he was renowned as a mathematician[11]. His name may be derived from Italian Doria. He translated Heron and Diophantos from Greek into Latin and edited new translations[12] of the books by Autolycus (*De Sphaera*[13]), Theodosius of Tripoli[14] and Euclid (*Phaenomena*[15]).

For his Diophantos edition, Auria wrote a number of preparatory manuscripts[16]. In four of these texts, we encounter notes on omissions, and two also contain his Latin translation[17]. In the Parisian manuscript, the Greek text appears on the left side, the Latin translation on the right. Corrections were added in the margins, with reference to other codices. Auria's translation makes references to or emends Xylander's translation. The manuscript also includes the book on polygonal numbers, but here the right-hand side –presumably intended for the translation– has remained blank. Further contained in this volume is Auria's Latin translation of Heron's *Automatopoetica*.

---

[11] J.J. HOFMAN(1698), lemma JOSEPHUS Auria.

[12] G. JÖCHER(1960-61) I, p.662.

[13] Joseph AURIA (ed.) Autolycus, *De sphera quae movetur liber*, Theodosii Tripolitae, *De habitationibus liber; Omnia scholijs antiquis & figuris illustrata; de Vaticana bibliotheca deprompta: & nunc primum in lucem edita. Josepho Auria. Neapol. Interprete. His additae sunt Maurolyci annotationes*, Apud haeredes Antonij Bladij, Rome, 1587.

[14] A.J. AURIA (ed;), *Theodosius Tripol., De diebus et noctibus in linguam latinam conversi A.J. Auria* [s.l.][s.n.], 1591.

[15] Joseph AURIA (ed.), *Euclides Phaenomenae Post Zamberti: et Maurolyci editionem, nunc tandem de Vaticana, Bibliotheca deprompta ... et de Graeca lingua in Latinam conuersa. A Iosepho. Auria Neapolitano. His additae sunt Maurolyci ... annotationes...*, Giovanni Martinelli, Rome, 1591.

[16] A. ALLARD(1981a), pp.104-107.

[17] Parisinus gr. 2380 and Ambrosianus E5inf, see A. ALLARD (1981a).

The manuscript Ambrosianus E5inf is Auria's own manuscript. In this version, we find a list of abbreviations, notes on the plus and minus signs, an alphabetic list of Greek numerals and their Hindu-Arab and Roman counterparts, as well as the *Book on polygonal numbers*, the *Arithmetika* with the *Book on polygonal numbers* in a Latin translation, and with references to and emendations of Xylander's translation. This manuscript is based on manuscripts of the Planudean and the non-Planudean class[18].

Auria's edition could have become a synthesis of the two classes of Diophantos manuscripts, alongside the emendations by the humanists. The undertaking was nearly successful, but unfortunately the manuscript never got to print. Moreover, Auria was not particularly consistent and rigorous in his reading and solving of certain difficulties. Remarkably given his reputation, this was more often than not due to a lack of mathematical intuition. Some of the emendations that Tannery attributes to Auria may in fact have been made by some of his Humanist colleagues, and the circle around Gian-Vincenzo Pinelli in particular.

It was not until 1621 that a new edition of Diophantos saw the light of day. It would be the single most influential edition, not in the least because it caught the imagination of mathematicians. Mathematicians were now able to delve deeper into the number theoretical consequences of many of the Diophantine problems. The edition was published by Claude-Gaspar Bachet de Méziriac (1581-1638)[19].

Claude-Gaspar was born on 9 October 1581 in Bourg-en-Bresse, the son of Jehan, a judge and counsellor to the Duke of Savoye, and Marie-Françoise de Chavanes. Jehan and Marie-Françoise had at least six children. A year after Marie-Françoise's death in 1586, Jehan remarried, but he died shortly after from the plague. Claude-Gaspar is assumed to have been educated at a Jesuit college, although direct evidence is lacking. He is known to have travelled to Rome and Paris among other places, but dates are lacking, with the exception of a stay in Paris in 1619-20. During his time in Paris, he had a number of his manuscripts published. After his sojourn in Italy, Bachet lived in his hometown and in his country mansion in the nearby commune of Jasseron. It was here, in these familiar surroundings, that he prepared his publications. He read, consulted and cited a great many authors, from which we may infer that he had access to an extensive library, most probably his own. In 1612, he had his first book published: *Problèmes plaisants et délectables qui se font par des nombres* (Lyons, 1612).

He married Philiberte de Chabeu in 1621, when he was already forty. The couple would be blessed with seven children.

Bachet was one of the members to be admitted to the august body of the *Académie Française* when it was established in 1634. He published not only mathematical treatises, but also literary work, including Ovid, and his own version of *Aesopus*. . .

---

[18]The 'descent' of Auria's manuscript has been described by A. ALLARD(1981a).

[19]For a biography of Bachet see C.G. COLLET & J. ITARD(1947).

Figure 8.1 *Bachet's Diophantos, Drouart, Lutetiae, 1621. Erfgoedbibliotheek Hendrik Conscience, Antwerpen, G 4803.*

For his edition of the Greek version of Diophantos' *Arithmetika*, Bachet relied primarily on the manuscript Parisinus 2379(r)[20]. André Allard was able to verify that all fillings of lacunae by Bachet also appear in that manuscript. He further made use of a partial copy of a Vatican manuscript produced by Jacques Sirmond (1559-1651)[21] and of notes that Claude Saumaise (1588-1653)[22] made from a manuscript by Andras Dudith. It goes without saying that he also had at his disposal earlier printed versions, such as those by Bombelli and Stevin[23].

Bachet's edition of the *Arithmetika* is preceded by three books of *Porisms* containing resp. 24, 21 and 19 theorems. These theorems are, on the one hand, an attempt to reconstruct Diophantos' *Porisms* and, on the other, lemmas to support the solutions of Diophantos' problems. They were written in a style reminiscent of Euclid's arithmetical books.

In Bachet's *Arithmetika*, each page is divided into two columns: the left column contains the Latin translation, the right the original Greek text. Bachet's comments are added in full. They sometimes provide an explanation of Diophantos'

---

[20]See A. ALLARD(1982-83), p.131. See also the stemma in appendix.

[21]Secretary to Superior General Aquaviva from 1590 to 1608.

[22]French humanist and philologer who wrote over eighty books. Successor to Joseph Scaliger at Leyden University (1631).

[23]C.G. COLLET & J. ITARD(1947), p.37.

method or propose new theorems. For example, in a comment on IV.31(p. 240-242), Bachet notes that any number can be written as the sum of at most four squares. He then provides a table illustrating this proposition[24].

Like all translators and commentators, Bachet considered the *Arithmetika* to be algebra, noting that the terminology was essentially algebraic. His identification of the *Arithmetika* with algebra led him to study the transformation of classical algebra into cossic algebra. For generalizations, Bachet relied on the same techniques as his predecessors. Moreover, he used them very consistently to find new solutions.

Because he was so apt at applying classical knowledge, his work illustrates one of the great dilemmas of the humanist project: how true should one remain to the original? In fact, what Bachet did was to gain an understanding of Diophantos in order to formulate new theorems that were classical in style, but new in content. More than once, he succeeded in arriving at more general solutions than those offered in the *Arithmetika*.

Bachet not only provided new solutions, he also developed new techniques to arrive at those solutions, and he even added theorems. In this sense, he not only emulated the classical authors, but actually surpassed them.

In his historical introduction, Bachet struggles with the same question facing every editor before and after him: "When did Diophantos live?". He concluded this had to be between Hypsikles and Hypatia, but was unable to put forward a more precise date[25]. To him, Diophantos could not but have been the father of algebra, considering that he lived long before Arab algebra came to fruition[26].

If the *Arithmetika* is algebra, then the book must contain the rules of this art or Diophantos must at least have been familiar with them. On this basis, Bachet interprets the sentence 'as is clear to see', which accompanies a condition, as an indication of Diophantos's awareness of these rules.

For example, from I.30 he infers that the rules for solving a quadratic equation must have been known:

> *Find two numbers whose difference and product are two given numbers.*

> It is necessary that the quadruple product of the numbers added to the square of their difference is a square, as is clear to see.

To us, this condition says no more than that the discriminant of the resulting quadratic equation must be positive[27].

---

[24]Upon reading this paragraph, Fermat noted that any number can be written as a sum of at most n n-gon numbers. This was proved by Lagrange, Legendre, Gauss for squares ($n = 4$) and generally by Cauchy.

[25]C.-G. BACHET(1621), p.iii.

[26]C.-G. BACHET (1621), pp.iii-iv.

[27] $\begin{cases} x - y & = & V \\ xy & = & P \end{cases}$ leads to the quadratic equation $x^2 - Vx - P = 0$ with discriminant $D = V^2 - 4P$.

Bachet not only identified the *Arithmetika* with algebra, he also assumed that algebra belonged to the realm of arithmetic. To Bachet, the *Arithmetika* was the art of numbers, not of quantities or geometrical entities. Yet there was little in the *Arithmetika* that linked it to classical number theory. For example, Diophantos allows fractions, whereas classical number theory only allows positive integers larger than 1.

Bachet introduced new methods that could be applied to integers. In more than one comment, he looks for ways to reduce the Diophantine solutions to integers and to find integer solutions using Diophantine methods. Bachet tried to incorporate these new problems into a tradition, albeit that of merchants' algebra. When solving an indeterminate equation, he sometimes asks for an integer solution that refers to the quality of the unknown, whereby the unknowns may refer to indivisible entities, such as animals or humans. This reference immediately implies that Bachet linked the problem to merchants' algebra and considered it to be an integral part of classical number theory. Unfortunately for Bachet, not all indeterminate equations have integer solutions though.

We are inclined, however, to situate the origins of number theory as we know it in the work of Bachet. The added problems to the *Arithmetika* are of such a general nature that they can no longer be seen as being part of recreational mathematics, nor are they specific enough to be part of merchants' algebra.

Consider the following example:

> *Given two numbers that are relatively prime, find a multiple of the first, which exceeds a multiple of the second with a given number in such a way that these multiples are as small as possible*[28].

Unlike Bombelli and Viète, Bachet remained faithful to the classical style, but evidently went much further in content, so that we can safely say he was a research mathematician of great stature. His preconditions for working with numbers made him shy away from Viète's algebra of kinds.

With Bachet's edition of the *Arithmetika*, Diophantine analysis enters the stage of contemporary mathematics. Ironically, it would be a zealous student of Viète who would find in it a sheer inexhaustible source of inspiration. His name was Pierre de Fermat.

---

[28]For example, if a, b and a difference c are given numbers, find m and n such that $ma - nb = c$ in such a way that if $m_1 a - n_1 b = c$ then $m < m_1$ and $n < n_1$.

8.3   This margin is too small...

Born either in August 1601 or in 1607[29], Pierre de Fermat was the son of Dominique de Fermat, a leather salesman, and Claire de Long, a descendant of a
family of lawyers. Pierre would follow in their footsteps. He bought the office of
conseiller au Parlement de Toulouse et commissaire aux requêtes du Palais. Although he had admirable administrative and legal skills, and possessed extensive
philological knowledge, we shall focus exclusively on his no less than extraordinary
contribution to mathematics.

Fermat had little inducement to publish. Like Viète, he was "merely" an amateur
mathematician. It would not be until after his death that much of his work was
published, by his son among others. Fermat had a great respect for the classical
authors, unlike his contemporary Descartes. Yet, ironically, by trying to renew
or continue classical traditions, he gave new directions to mathematical research,
which diverged ever further from the classics.

His first mathematical work was an attempt to translate Apollonios' treatise on
plane loci. This led him to formulate the same principle as Descartes:

> *Whenever two unknown quantities are found in final equality, there re-*
> *sults a locus [fixed] in place, and the endpoint of these [unknown quan-*
> *tities] describes a straight line or a curve*[30].

Fermat came to this conclusion through the algebraic methods, pioneered
by his mentor Viète (and his pupil Marino Ghetaldi), for solving the Apollonian
problems. In his manuscript *Ad locus planos et solidos isagoge*, Fermat reaches the
conclusion that a first-degree equation represents a straight line. He subsequently
studies second-degree curves and is able to reduce these, by translating or rotating
the system of axis to an equation of an ellipse, parabola or hyperbola.

Unfortunately his treatise was not published until 1679, about forty years after
Descartes' publication. Yet there is an essential difference between the two theories: Descartes started out with a curve and found the equation, whereas Fermat
started with the equation and found a curve. Or how, independently of each other,
two Frenchmen studied different sides of the same coin.

---

[29]On Fermat's life and mathematical work, see the seminal M.S. MAHONEY(1994). The
following paragraphs are in large part based on this book.

It has always been held that Fermat was born in 1601. However, Klaus Barner (2001), on the basis
of two documents recently discovered in the archives of Montauban and Beaumont-de-Lomagne,
comes to the conclusion that his birthdate must have been 1607. They show that Pierre de
Fermat's father Dominique was married twice and that the child Pierre, who was baptized on
August 20, 1601, is the son of his first wife Françoise Cazeneuve, whereas Pierre de Fermat's
mother is Claire de Long, Dominique's second wife. The true year of birth 1607/(08) is hinted at
by the last line of the epitaph above Fermat's tomb, according to which he died on 12 January
1665 at the age of 57.

[30]Translation by M.S. MAHONEY(1973), p.78.

Fermat's significance to mathematics is not limited to analytic geometry. He also made interesting contributions to proto-infinitesimal calculus. In 1637, he wrote a manuscript entitled *Methodus ad disquirendam maximum et minimum*, in which he tries to find the maxima and minima of certain functions. His idea was that values of a continuous function only differ very slightly if they are in one another's vicinity.

Fermat's most important contribution to mathematics, though, relates to number theory. This field had most probably come to his attention through Bachet's edition of the *Arithmetika*.

Fermat's interpretation of the problems is truer to Diophantos than Bachet's, except that Fermat allows only integers as solutions. In this way, he detaches Diophantos from algebra, where his predecessors had put him, and puts him in the realm of number theory. His contribution was initially met with little enthusiasm. Fermat was, after all, not a member of the mathematical fraternity and the new direction he was following did not appeal to those who were. More so than in any other branch of mathematics he was involved in, he remained very secretive about his findings in relation to number theory.

Many of his contributions appear as "marginal notes" in his copy of Bachet's Diophantos. He provides hardly any proofs. We may assume him to have been able to produce these proofs – or at least to believe he could – but that he always kept them to himself. Every now and again, he provides a glimpse of a proof in his correspondence. Only in his letters to Jacques de Billy s.j.(1602-1679)[31] did he reveal his improved methods for solving double equations.

It was only after his father's death that Samuel de Fermat published Pierre's mathematical work. To do so, he had to assemble the dispersed notes and try to establish some kind of order in them. The first book he had published was a re-edition of Bachet's *Arithmetika*, to which he added his father's notes. Jacques de Billy s.j. wrote an appendix, entitled *Doctrinae Analyticae Inventum Novum*. It was based on letters Fermat had written to him and in which he had explained the method for solving Diophantine equations.

In 1643, Fermat presented three problems to Pierre Brûlart de Saint Martin, with whom he maintained a correspondence on a number topics. One of his other correspondents on these particular queries was Bernard Frenicle de

---

[31] Jacques de Billy was a Jesuit and throughout his life taught mathematics at the colleges of Reims, Grenoble and Dijon. He became rector of the college at Châlons, Langres and Sens. He was befriended to Bachet and corresponded with Fermat. In de Billy Fermat found a confidant whom he trusted and confided some of his proofs to. De Billy published astronomical tables and made some progress in number theory.

Bessy[32] (1605-1675). The problems went as follows:

1. find a right-angled triangle such that the hypotenuse is a square and the sum of the perpendiculars, or of all three sides, is also a square[33].

2. find four right-angled triangles having the same area[34].

3. find a right-angled triangle such that the area plus the square of the sum of the smaller sides is a square.

Brûlart and Frenicle accused Fermat of having posed impossible problems. Fermat conceded to father Mersenne that the problems were extremely hard to solve and in the end he provided his correspondents with the answers. Fermat had found a method for solving double equations and used it to pose and prove new problems. What this boiled down to was that, once one has obtained a solution, one is able to construct an infinitude of solutions[35]. Fermat had unlocked the secret behind Diophantos' solutions, rendering his problems uninteresting. He now turned his attention to the divisibility of integers and the role of prime numbers. Fermat formulates propositions such as: "Any number can be written as the sum of at most four squares". His researches into the properties of square numbers brought him to the brink of Gauss's quadratic forms theory. He dedicated much of his time to equations of the type $x^2 - py^2 = \pm 1$, in which p is no square and x and y are integers.

Some of the theorems we owe to Fermat – without proof – are

- *If p is prime and p is a divisor of a then $a^{p-1} - 1$ is divisible by p*

- *Any uneven prime number can be written in a unique way as a difference of two squares.*

- *A prime number of the form $4n + 1$ can be written as the sum of two squares. Moreover it can only be the hypotenuse of a unique right triangle, the square can be a hypotenuse twice, the cube thrice, the fourth power four times etc.*

---

[32] Frenicle de Bessy was an excellent amateur mathematician who held an official position as a counsellor at the Court of Monnais in Paris. He corresponded with Descartes, Fermat, Huygens and Mersenne, mostly, but not exclusively on number theory. He solved many of the problems posed by Fermat introducing new ideas and posing further questions. See J.J. O'CONNOR and E F ROBERTSON(2000). On these problems see M.S. MAHONEY(1973), pp.307ff.

[33] E. BRASSINE(1853), pp.125-126.

[34] E. BRASSINE(1853), pp. 92-94. In his copy of Diophantos Fermat noticed that an infinitude of such triangles can be found.

[35] Fermat wrote a treatise on this topic, which has unfortunately been lost. Undoubtedly the gist of it is contained in de Billy's *Inventum Novum*. Consider the system $\begin{cases} f(x) & = & y^2 \\ g(x) & = & z^2 \end{cases}$ and suppose $x = m$ is a solution i.e. $\begin{cases} f(m) & = & s^2 \\ g(m) & = & t^2 \end{cases}$. Now substitute x by $u + m$ to obtain $\begin{cases} f(u+m) = F(u) & = & y^2 \\ g(u+m) = G(u) & = & z^2 \end{cases}$ the constant in $F(u)$ is s^2, while the constant in $G(u)$ is t^2, therefore the system is easily solved using Diophantine methods.

- *The equation $x^2 + 2 = y^3$ has only one solution; the equation $x^2 + 4 = y^2$ has two.*

- *There are no positive integers such that $x^4 - y^4 = z^2$*

- *There are no positive integers such that $x^n + y^n = z^n$*

The last proposition, known as Fermat's last theorem, was a marginal note to Diophantos' II.8. Unfortunately, the margin was too small to contain the proof:

> *Cubum autem in duos cubos, aut quadratoquadratorum in duos quadratos, et generaliter nullam in infinitum ultra quadratum potestatem in duos eiusdem nominis fas est dividere. Cubus rei demonstrationem mirabilim detexi hanc margines exiguitas non caperet.*

The theorem would torment the minds of the very best and brightest mathematicians for centuries to come. Fermat claimed to have an elegant proof – most probably based on his method of infinite descent and incorrect. He only gave the proof for $n = 4$. Euler proved the proposition for $n = 3$, Legendre and Dirichlet for $n = 5$ and Lamé for $n = 7$. Kummer once thought he had found the proof, but this was not the case. In fact, it has been suggested that this is the theorem to have generated the largest number of faulty proofs.

Fermat's last theorem would not give up its secrets until 1994, when it was finally cracked by Andrew Wiles[36]. However, that is yet another chapter in Diophantine analysis...

---

[36] A. WILES(1995) and R. TAYLOR & A. WILES(1995). On the attempts to prove Fermat's Last Theorem see S. SINGH(1997) and A. ACZEL(1996).

Chapter 9

Coda: Hilbert's tenth problem

At the turn of the previous century, David Hilbert (1862-1943) was already regarded as one of the finest mathematicians of his generation[1]. He had forced
breakthroughs in the theory of invariants, number theory and geometry[2]. Hilbert
would, self-evidently, address the International Congress of Mathematics in 1900
on the occasion of the World's Fair in Paris. He conferred on the subject with his
friends Hermann Minkowski (1864-1909) and Adolf Hurwitz (1859-1919). They
advized him to look ahead towards the future. Organizationally, the Congress was
a disaster, remembered only for Hilbert's speech. A condensed version appeared in
L'Enseignement Mathématique (1900). The complete version was published in the
Nachrichten (1900) of the Göttinger Wissenschaftsgesellschaft and the following
year also in *Archiv der Mathematik und Physik*[3]. The problems Hilbert proposed,
and which he felt needed resolving, came from all fields of mathematics. They dealt
with divergent aspects such as the cardinality of natural and real numbers, the
axiomatization of mathematics and physics, algebraic number theory, geometry,
algebra, and analysis.

The tenth problem goes as follows[4]:

> *Entscheidung der Lösbarkeit einer diophantische Gleichung.*
> (Determination of the solvability of a diophantine equation)

> Eine Diophantische Gleichung mit irgend welchen Unbekannten und mit
> ganzen rationalen Zahlencoefficienten sei vorgelegt: man soll ein Ver-
> fahren angeben, nach welchem sich mittelst einer endlichen Anzahl von

---

[1] On David Hilbert and the problems he proposed see: I. GRATTAN-GUINNESS (2000), J.J.
GRAY (2000).
[2] *Grundlagen der Geometrie* (1899) contains a new and complete axiomatization of the Euclidean geometry.
[3] D. HILBERT(1900).
[4] On this problem, see M. DAVIS & R. HERSH(1973).

Operationen entscheiden läszt, ob die Gleichung in ganzen rationalen Zahlen lösbar ist.

The answer to this problem is affirmative for polynomials in one unknown, because it is possible to determine a lower and an upper bound for the possible solutions in terms of the coefficients. The answer to the general problem, however, is negative, as was subsequently proved by Youri Matijasevitch. The proof was read to the symposium on Hilbert's Problems of the American Mathematical Society in 1974. It rests on the fact that there exists a polynomial expression in thirteen unknowns and with one parameter for which there is no algorithm that shows whether the equation has solutions for the given value of the parameter.

The proof is not only interesting as a solution to the tenth problem of Hilbert; it also has deep implications for number theory. For instance, it can now be proved that there is a polynomial whose non-negative integer values are exactly the prime numbers.

Goldbach's conjecture may be reduced to a Diophantine equation. Had Hilbert's tenth problem been confirmed, then Goldbach's conjecture would have been false. The Riemann hypothesis, too, can be reduced to a Diophantine equation, which does not actually make it easier to solve.

The most intriguing consequence, however, is that every mathematical theory comes with at least one Diophantine equation that has no solutions if and only if the theory is consistent.

Chapter 10

Stemma

Known manucripts with sigels as assigned by P. Tannery and A. Allard. For a detailed description of these manuscripts we refer to the work of A. Allard. Stemma of Diophantosmanuscripts, see A. Allard (1980), (1981a), (1981b), (1982-83), (1983), (1984), (1988).

a. Milan, *Biblioteca Ambrosiana A91 sup.*, parchment, fifteenth century.

E. Milan, *Biblioteca Ambrosiana E 5 inf.*, paper, sixteenth century.

M. Milan, *Biblioteca Ambrosiana Et 157 sup.*, paper, thirteenth century. (Copy of Maximos Planudes. Archetype of the Planudean manuscripts.)

o. Milan, *Biblioteca Ambrosiana Q 121 sup.*, paper, sixteenth century.

d. Krakau, *Bibliotheka Jagiellonska 544*, paper, sixteenth century.

h. Wolfenbüttel, *Gudianus gr. 1*, paper, sixteenth century.

hII. Fragment containing problem I.1, part of of h.

q. Firenze, *Biblioteca Laurentianus Acquisti e Doni 163-164*, paper, sixteenth century.

B. Venice, *Biblioteca Marciana gr. 308*, paper, thirteenth century (ff. 1-49 written on 16th century paper, 50-254 on 13th century paper).

A. Madrid, *Biblioteca Nacional 4678*, paper, thirteenth century.

n. Naples, *Borbonicus III C17*, paper, sixteenth century.

f. Oxford, *Baroccianus 166*, paper, sixteenth century.

P. Vatican City, *Biblioteca Vaticana Palatinus gr. 391*, paper, sixteenth century.

O. Oxford Bodleian, *Savilianus 6*, paper, sixteenth century.

A. Meskens, *Travelling Mathematics - The Fate of Diophantos' Arithmetic*, Science Networks. Historical Studies 41, DOI 10.1007/978-3-0346-0643-1_10, © Springer Basel AG 2010

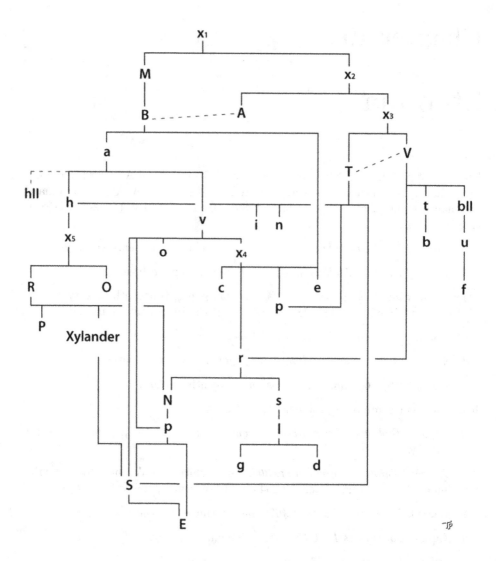

s. Paris, *Bibliothèque nationale Arsenacensis 8406*, paper, sixteenth century.

i. Paris, *Bibliothèque nationale gr. 2378*, paper, sixteenth century.

r. Paris, *Bibliothèque nationale gr. 2379*, paper, sixteenth century.

S. Paris, *Bibliothèque nationale gr. 2380*, paper, sixteenth century.

p. Paris, *Bibliothèque nationale gr. 2485*, paper, sixteenth century.

g. Madrid, *Escorial R II 3*, paper, sixteenth century.

c. Madrid, *Escorial R III 18*, paper, sixteenth century.

e. Madrid, *Escorial T I 11*, paper, sixteenth century.

l. Madrid, *Escorial I 15*, paper, sixteenth century.

N. Turin, *Biblioteca nazionale C I 4*, paper, sixteenth century.

b. Vatican City, *Biblioteca Vaticana Urbinas Universitatis 102*, paper, sixteenth century.

bII. Fragments of book I, part of b.

u. Vatican City, *Biblioteca Vaticana Barberinianus gr. 267*, paper, sixteenth century.

V. Vatican City, *Biblioteca Vaticana gr. 191*, paper, thirteenth century.

v. Vatican City, *Biblioteca Vaticana gr. 200*, paper, fifteenth century.

T. Vatican City, *Biblioteca Vaticana gr. 304*, paper, fourteenth century.

R. Vatican City, *Biblioteca Vaticana Reginensis gr. 128*, paper, sixteenth century.

t. Vatican City, *Biblioteca Vaticana Urbinas gr. 74*, paper, sixteenth century.

Manuscripts not mentioned by Allard, but which can be found in Jordanus database:

- Milan, *Biblioteca Ambrosiana C263 Inf.* Greek.
  Many treatises among which

    - Diophantus, *Prolegomena in Almagestum*

    - Diophantus, *Opera quadeam de mathematicis*

  Jordanus IMILAC263I

- Leiden, *Universiteitsbibliotheek B.P.G. 74 G.* Greek.

    - Diophantus, *Libri arithmetici* (excerpt.), *Prolegomena arithmetica, Arithmeticis excerptum*. Incipit: *Ek ton tu Diophantu arithmetickon: apo pantos arithmu tetragonu monados.*

  Jordanus NLEIUBPG074G

Bibliography

[1] *Simon Stevin 1548-1620; De geboorte van een nieuwe wetenschap.* Brepols-KBR, 2004.

[2] A. Aczel, *Fermat's Last Theorem.* Four Walls Eight Windows, 1996.

[3] A. Agostini, *Un commento su Diofanto contenuto nel MSS Palat.625.* Archeion **11**(1929), 41-53.

[4] C. Alexandre, *Dictionnaire Grec-Français.* Hachette, 1888.

[5] A. Allard, *L'Ambrosianus Et157 sup. Un manuscrit autographe de Maxime Planude.* Scriptorium **33**(1979), 219-234.

[6] A. Allard, *Diophante d'Alexandrie, les Arithmétiques: histoire du texte grec, éditions critique, traductions, scolies.* FNRS, 1980.

[7] A. Allard, *La tentative d'édition des Arithmétiques de Diophante d'Alexandrie.* Revue d'histoire des textes **11**(1981a), 99-122.

[8] A. Allard, *Maxime Planude: Le grand calcul selon les Indiens.* Travaux de la Faculté de Philosophie et Lettres de l'Université Catholique de Louvain, 1981b.

[9] A. Allard, *La tradition du texte grec des Arithmétiques de Diophante d'Alexandrie.* Revue d'histoire des textes **12-13**(1982-83), 57-137.

[10] A. Allard, *Les scolies aux arithmétiques de Diophante d'Alexandrie dans le Matritensis Bibl. Nat. 4678 et les Vaticani gr. 191 et 304.* Byzantion, Revue int. des études byzantines **53**(1983), 664-760.

[11] A. Allard, *Un exemple de transmission d'un texte grec scientifique: le Mediolanensis Ambrosianus A91 sup., un manuscrit de Jean Vincent Pinelli prêté à Mathieu Macigno.* Les Études Classiques **52**(1984), 317-331.

[12] A. Allard, *A propos de plusieurs publications récentes.* Revue des Questions Scientifiques **CLV**(1984), 309-316 & **CLVIII**(1987), 375-384.

[13] A. Allard, *Le manuscrit des arithmétiques de Diophante d'Alexandrie et les lettres d'Andrè Dudith dans le Monacensis Lat 10370.* in: M. Folkerts, U. Lindgren(1985), p.297-315.

[14] C. Alvarez J., *François Viète et la mise en équation des problèmes solides.* in: P. Radelet-de Grave (2008), 17-61.

[15] L. Ambjörn, *Qusta ibn Luqa On Numbness. A Book on Numbness, its Kinds, Causes and Treatment according to the opinion of Galen and Hippocrates. Edition, translation and commentary.* Studia Orientalia Lundensia Nova Series vol. **1**, 2000.

[16] A. Anbouba, *L' algèbre arabe aux IX et X siècle.* J. for the History of Arabic Science **2**/**1**(1978).

[17] A. Anbouba, *Un traité d'Abū Ja'far [al Khāzin] sur les triangles rectangles numériques.* J. for the History of Arabic Science **3**/**1**(1979), 134-177.

[18] M. Angold, *The Byzantine Empire.* Longman, 1997.

[19] G. Argoud, *Science et vie intellectuelle à Alexandrie.* Université de St-Etienne, 1994.

[20] G. Argoud, *Héron d'Alexandrie, mathématicien et inventeur.* in: G. Argoud (1994), 53-66.

[21] G. Argoud & J.Y. Guillaumin, *Sciences exactes et sciences appliquées à Alexandrie.* Université de St-Etienne, 1998.

[22] M. Asper, *Dionysius (Heron, Def. 14.3) und die Datierung Herons von Alexandria.* Hermes **129**(2001), 135-137.

[23] Athenaeus, C.D. Yonge (ed.), *The deipnosophists, or, Banquet of the learned of Athenœus* I. Henry G. Bohn, s.d. (digital.library.wisc.edu/1711.dl/Literature.AthV1)

[24] C.-G. Bachet de Méziriac, *Diophanti Alexandrini Arithmeticorum.* Drouart, 1621.

[25] R. Bagnall, *Egypt in Late Antiquity.* Princeton UP, 1993.

[26] R. Bagnall, *Alexandria: Library of Dreams.* Proc. of the American Phil. Soc. **146**(2002), 348-362.

[27] E. Barbin (ed.), *François Viète; Un mathèmaticien sous la Renaissance.* Vuibert, 2005.

[28] K. Barner, *How Old did Fermat become?.* NTM Int. J. of History & Ethics of Natural Sciences, Technology & Medicine **9**(2001), 209-228.

[29] K. Barner, *Negative Grössen bei Diophant I & II.* NTM Int. J. of History & Ethics of Natural Sciences, Technology & Medicine **15**(2007), 18-49 & 98-117.

[30] M. Bartolozzi & R. Franca, *La teoria delle proporzioni nella matematica dell'abaco da Leonardo Pisano a Luca Pacioli.* Bolletino di Storia delle Scienze Matematiche **10**(1990), 3-28.

[31] I. Bashmakova, *Diophante et Fermat*. Revue d'Histoire des Sciences **19** (1966), 289-306.

[32] I.G. Bashmakova & E.I. Slavutin, *'Genesis triangulorum' de François Viète et ses récherches dans l'analyse indeterminée*. Arch. for the History of Exact Sciences **16**(1976/77), 289-306.

[33] I.G. Bashmakova, *Arithmetica of Algebraic Curves from Diophantus to Poincaré*. Historia Mathematica **8**(1981), 393-416.

[34] I.G. Bashmakova, *Diophantus and Diophantine Equations*. Math. Ass. of America, 1997 (translation of the Russian edition 1972).

[35] I.G. Bashmakova & G. Smirnova, *The Beginnings and Evolution of Algebra*. Math. Ass. of America, 2000 (translation of the Russian edition).

[36] P. Benoît & F. Micheau, *The Arab Intermediary*. in: M. Serres(1995), 191-221.

[37] J.L. Berggren, *Greek and Islamic Elements in Arabic Mathematics*. in: I. Müller(1991), 195-217.

[38] J.L. Berggren, *Islamic acquisition of the foreign sciences: a cultural perspective* in: F. J. Ragep(1996), 263-283.

[39] A. Birkenmajer, *Diophante et Euclide*. Studia Copernicana **1**(1970), 575-585.

[40] P. Bockstaele, *Gielis van den Hoecke en zijn* Sonderlinghe boeck in dye edel Conste Arithmetica. Academia Analecta **47/1**(1985), 1-29.

[41] R. Bombelli, E. Bortolotti (ed.), *L'Algebra, Opera di Rafael Bombelli da Bologna*. Feltrinelli, 1966.

[42] B. Boncompagni, *Recherches sur plusieurs ouvrages de Leonard de Pise*. Imprimerie des Sciences mathématiques et physiques, 1851.

[43] H.J.M. Bos, *Redefining Geometrical Exactness*. Springer, 2001.

[44] H.J.M. Bos, *Simon Stevin, wiskundige*. in: Simon Stevin(2004), 49-62.

[45] H. Bosmans, *Le* De arte magna *de Guillaume Gosselin?*. Bibliotheca Mathematica 3.Folge **7/1**(1906), 44-66.

[46] H. Bosmans, *Diophante d'Alexandrie*. Revue des Questions Scientifiques (1926), 443-456.

[47] H. Botermann, *Das Judenedikt des Kaiser Claudius*. Steiner, 1996.

[48] A.K. Bowman, *Egypt after the Pharaohs 332 BC-AD 642, from Alexander to the Arab Conquest*. British Museum, 1986.

[49] E. Brassine, *Précis des oeuvres mathématiques de Fermat et de l'Arithmétique de Diophante*. Douladouré, 1853.

[50] N. Bubnov, *Gerberti postea Silvestri II papae. Opera Mathematica (972-1003)*. Georg Olms Verlag, 1963.

[51] L.N.H. Bunt, S. Jones, J.D. Bedient, *The Historical Roots of Elementary Mathematics*. Dover, 1976.

[52] D. Burton, *History Of Mathematics*. Brown, 1991-95.

[53] J.S. Byrne, *A Humanist History of Mathematics? Regiomontanus' Padua Oration in Context*. J. of the History of Ideas **67**(2006), 41-61.

[54] F. Cajori, *A History of Mathematical Notations*. Dover, 1993.

[55] A. Cameron, *Isidore of Miletus and Hypatia: on the Editing of Mathematical Text*. Greek, Roman and Byzantine Studies **31**(1990), 103-127.

[56] L. Canfora, J.P. Monganaro & D. Dubroca, *La véritable histoire de la bibliothèque d'Alexandrie*. Les chemins d'Italie - Dejonquères, 1988.

[57] Cassius Dio, *Roman History*. The Loeb Classical Library, 1927. (penelope.uchicago.edu/Thayer/E/Roman/home.html)

[58] G. Cavallo, *Libri scritture scribe a Ercolano*. Cronache Ercolanesi **13** Suppl 1, 1983.

[59] M. Caveing, *La figure et le nombre. Récherches sur les premières mathématiques de Grecs*. Université de Lille III, 1982.

[60] R. Cavenaille, *Pour une histoire politique et sociale d'Alexandrie. Les origines*. L'Antiquité Classique **XLI**(1972), 94-112.

[61] R. Cessi, *Bartolomeo Camillo Zanetti, tipografi e calligrafi del '500*. Archivio Veneto-Tridentino **8**(1925), 174-182.

[62] K. Chemla, A. Allard & R. Morelon, *La tradition arabe de Diophante d'Alexandrie*. L'Antiquité Classique **LV**(1986), 351-375.

[63] K. Chemla, *History of Science, History of Text*. Springer, 2005.

[64] S. Chrisomalis, *The Egyptian Origin of the Greek Alphabetic Numerals*. Antiquity **77**(2003), 485-496.

[65] S. Chrisomalis, *A Cognitive Typology for Numerical Notation*. Cambridge Archeological J. **14**(2004), 37-52.

[66] S. Chrisomalis, *Numerical Notation; a Comparative Study*. Cambridge UP, 2010.

[67] J. Christianides, *αριθμετικα στοιχησις: Un traité perdu de Diophante d'Alexandrie?*. Historia Mathematica **18**(1991), 239-246.

[68] J. Christianides, *On the History of Indeterminate Problems of the First Degree in Greek Mathematics*. in: C. Gavroglu(1994), 237-247.

[69] J. Christianides, *Maxime Planude sur le sens du terme diophantien 'plasmatikon'.* Historia Scientiarum **6**(1996), 37-41.

[70] J. Christianides, *Une Interpretation byzantine de Diophante.* Historia Mathematica **25**(1998), 22-28.

[71] J. Christianides, *Les lecteurs byzantins de Diophante.* in: E. Knobloch(2002), 153-163.

[72] J. Christianides, *Classics in the History of Greek Mathematics.* Kluwer Academic, 2004.

[73] G. Cifoletti, *Mathematics and Rhetoric.* UMI, 1992.

[74] G. Cifoletti, *The Creation of the History of Algebra in the sixteenth century.* in: C. Goldstein, J. Gray & J. Ritter(1996), 123-144.

[75] M.L. Clarke, *Higher Education in the Ancient World.* Routledge & Kegan Paul, 1971.

[76] W.E.H. Cockle, *Restoring and conserving papyri.* Bulletin of the Institute of Classical Studies **30**(1983), 147-165.

[77] C.G. Collet & J. Itard, *Un mathématicien humaniste: Claude-Gaspar Bachet de Méziriac.* Revue d'histoire des sciences **1**(1947), 26-50.

[78] C.N. Constantinides, *Higher Education in Byzantium in the Thirteenth and Early Fourteenth Centuries.* Cyprus Reasearch Centre, 1982.

[79] F.C. Conybeare, *Dionysius of Alexandria, Newly discovered letters to the Popes Stephen and Xystus.* English Historical Review **25**(1910), 111-114. (www.tertullian.org/fathers/dionysius_alexandria_letters.htm)

[80] F.C.Conybeare, *Timothy Aelurus, Patristic Testimonia.* J. of Theological Studies **15**(1914), 432-442. (www.tertullian.org/fathers/timothy_aelurus_testimonia.htm).

[81] S. Corcoran, *The Praetorian prefect Modestus and Hero of Alexandria's Stereometrica.* Latomus **54**(1995), 377-384.

[82] P. Costil, *André Dudith: humaniste hongrois 1532-1589.* Belles Lettres, 1935.

[83] S. Couchoud, *Mathématique égyptiennes. Récherches sur les connaissances mathématiques de l'Egypte pharaonique.* Le Léopard d'Or, 1993.

[84] R. Cribiore, *Writing, teachers and students in Graeco-Roman Egypt.* Scholar Press, 1996.

[85] R. Cribiore, *Gymnastics of the Mind: Greek Education in Hellenistic and Roman Egypt.* Princeton UP, 2001.

[86] S. Cuomo, *Pappus of Alexandria and the Mathematics of Late Antiquity.* Cambridge UP, 2000.

[87] S. Cuomo, *Ancient mathematics*. Routledge, 2001.

[88] A. Czwalina, *Arithmetik des Diophantos aus Alexandria*. Vandenhoeck & Ruprecht, 1952.

[89] P. Damerow & W. Lefèvre, *Rechenstein, Experiment, Sprache Historische Fallstudien zur Entstehung der Exacten Wissenschaften*. Kett-Cotta, 1981.

[90] P. Damerow, *Kannten die Babylonier den Satz des Pythagoras?* in: J. Høyrup & P. Damerow(2001), 219-310.

[91] M. Davis & R. Hersh, *Hilbert's Tenth Problem*. Scientific American **229/5** (1973), 84-91.

[92] M.A.B. Deakin, *Hypatia and her Mathematics*. American Math. Monthly **101/3**(1994), 234-243.

[93] M.A.B. Deakin, *The primary sources for the Life and Work of Hypatia of Alexandria*. Monash University, 1995.

[94] M.A.B. Deakin, *Hypatia of Alexandria; mathematician and martyr*. Prometheus Books, 2007.

[95] D. Delia, *Alexandrian citizenship during the Roman Principate*. Scholars Press, 1991.

[96] D. Delia, *From Romance to Rhetoric: The Alexandrian Library in Classical and Islamic Traditions*. American Historical Review **97**(1992), 1449-1467.

[97] A. De Rosalia, *La vita de Constantino Lascaris*. Archivio Storico Siciliano III **9**(1957-58), 21-70.

[98] R. De Smet, *Simon Stevin en de paradox van het gefragmenteerde humanisme*. in: Simon Stevin(2004), 27-34.

[99] J.L. De Vaulezard, *La nouvelle algèbre de M. Viète*. Fayard, 1986.

[100] J. Dhombres, *Nombre, mesure et continu. Epistémologie et histoire*. CEDIC, 1980.

[101] L.E. Dickson, *History of the Theory of Numbers*. 3 vols. Chelsea, 1971.

[102] E.J. Dijksterhuis, *Simon Stevin*. Martinus Nijhoff, 1970.

[103] Y. Dold-Samplonius, et al. (eds.) *From China to Paris: 2000 Years Transmission of Mathematical Ideas*. Steiner, 2002.

[104] M. Dzielska, *Hypatia of Alexandria*. Harvard UP, 1996.

[105] E.L. Eisenstein, *The Printing Press as an Agent of Change. Communications and Cultural Transformations in Early Modern Europe*. 2 vols. Cambridge UP, 1979.

[106] M. El-Abbadi, *Life and fate of the ancient Library of Alexandria*. UNESCO, 1990/92.

[107] G. Endress & R. Kok, *The Ancient Tradition in Christian and Islamic Hellenism*. Research school CNWS, 1997.

[108] R.K. Englund, *Grain Accounting Practices in Archaic Mesopotamia*. in: J. Høyrup & P. Damerow(2001), 1-35.

[109] Epiphanius of Salamis, J.E. Dean (ed.), M. Spengling (foreword), *Weights and Measures, The Syriac Version*. University of Chicago Press, 1935. (www.tertullian.org/fathers)

[110] A. Erskine, *Culture and Power in Ptolomaic Egypt: The Museum and Library of Alexandria*. Greece and Rome **42**(1995), 38-48.

[111] Eusebius of Caesarea, Tr. E.H. Gifford (1903), *Praeparatio Evangelica (Preparation for the Gospel)*. (www.tertullian.org/fathers)

[112] Eusebius Pamphilius, P. Schaff (ed.), *Church History, Life of Constantine, Oration in Praise of Constantine*. Christian Literature Publishing Co., 1890. (www.ccel.org/ccel/schaff/npnf201.html and www.newadvent.org/fathers/)

[113] Eutropius, F. Ruehl (ed.), *Breviarium ab urbe condita*. B.G. Teubner Verlag, 1887. (www.forumromanum.org/literature/eutropius)

[114] G.R. Evans, *From Abacus to Algorism, Theory and Practice in Medieval Arithmetic*. British J. for the History of Science **10**(1977), 114-131.

[115] E. Evrard, *À quel titre Hypatie enseigna-t-elle la philosophie*. Revue des Études Grecques 90(1977), 69-74.

[116] D. Fideler (ed.), *Alexandria* **2**. Phanes Press, 1993.

[117] Flavius Josephus, William Whiston (trans.), *The Works of Flavius Josephus*. John E. Beardsley, 1895. (www.perseus.org)

[118] M. Folkerts, *Pseudo-Beda: De Arithmeticis proportionibus*. Südhoffs Archiv **56**(1972), 22-43.

[119] M. Folkerts, *Regiomontanus als Mathematiker*. Centaurus **21**(1977), 214-245.

[120] M. Folkerts, *Die mathematischen Studien Regiomontans in seiner Wiener Zeit*. in: G. Hamann (ed.), *Regiomontanus-Studien*, Österreichische Akademie der Wissenschaften, 1980, 175-209.

[121] M. Folkerts, *Regiomontanus als Vermittler algebraischen Wissens*. in: M. Folkerts, U. Lindgren(1985), 207-219.

[122] M. Folkerts, U. Lindgren, *Mathemata*. Franz Steiner Verlag, 1985.

[123] M. Folkerts, *Regiomontanus' role in the transmission and transformation of Greek Mathematics*. in: F.J. Ragep(1996), 89-113.

[124] M. Folkerts, *Regiomontanus' Role in the Transmission of Mathematical Problems.* in: Y. Dold-Samplonius, et al.(2002), 411-428.

[125] M. Folkerts, *Leonardo Fibonacci's Knowledge of Euclid's Elements and of other Mathematical Texts.* Bolletino di Storia delle Scienze Matematiche **24** (2004), 93-113.

[126] M. Folkerts, *The Development of Mathematics in medieval Europe: the Arabs, Euclid, Regiomontanus.* Variorum, 2005.

[127] M. Folkerts, *Die mathematischen Studien Regiomontans in seiner Wiener Zeit.* in: M. Folkerts(2005) (reprint of (1985)).

[128] D.H. Fowler, *The Mathematics of Plato's Academy; a New Reconstruction.* Clarendon Press, 1987.

[129] D.H. Fowler, *Dynamis, mithartum, and square.* Historia Mathematica **19**(1992a), 418-419.

[130] D.H. Fowler, *Ratio and proportion in early Greek mathematics.* in: A.C. Bowen(ed.), *Science and Philosophy in Classical Greece.* Garland, 1992b, 98-118.

[131] D.H. Fowler, *The story of the Discovery of Incommensurability, revisited.* in: C. Gavroglu(1994), 221-235.

[132] D.H. Fowler, *Logistic and Fractions in Early Greek Mathematics: a New Interpretation.* in: J. Christianides(2004), 367-380.

[133] R. Franci & L. Toti Rigatelli, *Towards a History of Algebra from Leonardo of Pisa to Luca Pacioli.* Janus **72**(1985), 17-82.

[134] D. Frankfurter, *Religion in Roman Egypt.* Princeton UP, 1998.

[135] P.M. Fraser, *Ptolomaic Alexandria.* Oxford UP, 1972.

[136] P. Freguglia, *Algebra e geometria in Viète.* Bolletino di Storia delle Scienze Matematiche **9**(1989), 49-90.

[137] P. Freguglia, *L'interprétation de l' œuvre de Diophante: Les Zeteticorum Libre Quinque.* in: E. Barbin(2005), 75-86.

[138] P. Freguglia, *Les équations algébriques et la géométrie chez les algébristes du XVIe siècle et chez Viète.* in: P. Radelet-de Grave(2008), 149-161.

[139] J. Friberg, *Traces of Babylonian influence in the Arithmetica of Diophantus.* Preprint of the Department of Mathematics, Chalmers University of Technology & Göteborg University **19**(1991).

[140] J. Friberg, *Excavation problems in Babylonian Mathematics.* SCIAMVS **4**(2003), 3-21.

[141] J. Friberg, *Unexpected Links between Egyptian and Babylonian Mathematics.* World Scientific, 2006.

[142] S. Gandz, *The Sources of al-Khwarizmi's Algebra.* Osiris **I**(1936), 236-277.

[143] C. Gavroglu, *Trends in the Historiography of Science.* Kluwer, 1994.

[144] A. Cornelius Gellius, *Noctes Atticae (Attic Nights).* The Loeb Classical Library, 1927. (penelope.uchicago.edu/Thayer/E/Roman/home.html)

[145] D. Geneakoplos, *Interactions of the 'Sibling' Byzantine and Western Cultures.* Yale UP, 1976.

[146] D. Geneakoplos, *Medieval Western Civilization and Byzantine and Islamic Worlds.* D.C. Heath, 1979.

[147] D. Geneakoplos, *Constantinople & the West: Essays on Late Byzantine (Palaeologan) and Italian Renaissances and the Byzantine & Roman Churches.* University of Wisconsin Press, 1984.

[148] H. Gerstinger & K. Vogel, *Eine Stereometrische Aufgabensammlung im Papyrus Graecus Vindobonensis 19996.* Nationalbibliothek Wien, 1932.

[149] M. Gigante, *Catalogo dei papiri Ercolanesi, Centro internazionale per lo studio dei papiri Ercolanesi.* Bibliopolis, 1979.

[150] J. Gill, *The Council of Florence.* Cambridge UP, 1961.

[151] C. Goldstein, J. Gray & J. Ritter, *L'Europe mathématique-Mathemetical Europe.* Editions de la Maison des Sciences de l'homme, 1996.

[152] G. Gosselin, *De Arte Magna.* Gilles Beys, 1577.

[153] P.W.G. Gordan, *Two Renaissance Book Hunters.* Colombia UP, 1974.

[154] R. Goulding, *Method and Mathematics: Petrus Ramus' Histories of the Sciences.* J. of the History of Ideas **67**(2006), 63-85.

[155] I. Grattan-Guinness, *A Sideways Look at Hilbert's Twenty-three Problems of 1900.* Notices of the AMS **47/7**(2000), 752-757.

[156] J.J. Gray, *The Hilbert Challenge.* Oxford UP, 2000.

[157] C. Haas, *Alexandria in Late Antiquity: topography and social conflict.* Johns Hopkins UP, 1997.

[158] R.W. Hadden, *On the Shoulders of Merchants.* State University of New York Press, 1994.

[159] H.V. Harris, G. Ruffini, *Ancient Alexandria between Egypt and Greece.* Brill, 2004.

[160] T.L. Heath, *Euclid. The Thirteen Books of the Elements.* 3 vols, Dover, 1956.

[161] T.L. Heath, *Diophantus of Alexandria, a Study in the History of Greek Algebra*. Dover, 1964.

[162] T.L. Heath, *A History of Greek Mathematics*. 2 vols. Dover, 1981.

[163] A. Heeffer, *The Methodological Relevance of the History of Mathematics for Mathematics Education*. preprint.
(logica.rug.ac.be/centrum/writings/pubs.php).

[164] Herodianus, M.F.A. Blok (trans.), *Caracalla in Alexandrië*. Hermeneus **57/3** (1985), 207-208.

[165] Herodotus, A. D. Godley (trans.), *Histories*. Harvard UP, 1920.
(www.perseus.org)

[166] D. Hilbert, *Mathematische Probleme; Vortrag gehalten auf dem internationalen Mathematiker-Kongress zu Paris*. Archiv für Mathematik und Physik, 1900.

[167] J. Hintikka & U. Remes, *The Method of Analysis: its Geometrical Origin and its General Significance*. D. Reidel, 1974.

[168] Hippolytus (Saint), J. H. MacMahon(trans.) *Refutation of All Heresies*. Christian Literature Publishing Co., 1886.
(www.newadvent.org/fathers/050101.htm)

[169] R. Hoche, *Hypathia, die Tochter Theons*. Philogus **15**(1860), 435-474.

[170] J.E. Hoffman, *Bombelli's Algebra -eine genialische Einzelleistung und Ihre Einwirkung auf Leibniz*. Studia Leibnitiana **4**(1972), 196-252.

[171] J.J. Hofmann, *Lexicon Universale*. Jacob Hackius et al., 1698.

[172] J.P. Hogendijk, *Transmission, transformation and originality: the relation of Arabic to Greek Geometry*. in F.J. Ragep(1996), 31-64.

[173] J. Høyrup, *Dynamis, the Babylonians and Theaetetus 147c7-148d7*. Historia Mathematica **17**(1990), 201-222.

[174] J. Høyrup, *Sub-Scientific Mathematics. Observations on a Pre-Modern Phenomenon*. History of Science **28**(1990b), 63-86.

[175] J. Høyrup, *In Measure, Number and Weight*. State University of New York Press, 1994.

[176] J. Høyrup, *The formation of a Myth: Greek Mathematics - our mathematics*. in: C. Goldstein, J. Gray & J. Ritter(1996), 103-122.

[177] J. Høyrup, *Hero, Ps-Hero and Near Eastern Practical Geometry. An Investigation of Metrica, Geometrica and other Treatises*. 3 Række: Preprints og reprints, 1996.

[178] J. Høyrup, *Alchemy and mathematics, Technical knowledge subservient to ancient γνωσις.* 3 Række: Preprints og reprints, 2000a.

[179] J. Høyrup, Hero, *Seleucid innovations in the Babylonian "Algebraic" Tradition and their Kin abroad.* 3 Række: Preprints og reprints, Roskilde, 2000b.

[180] J. Høyrup, *On a Collection of Geometrical Riddles and their Rôle in the shaping of 4 to 6 "Algebras".* Science in Context **14**(2001), 85-131.

[181] J. Høyrup & P. Damerow, eds., *Changing Views on Ancient Near Eastern Mathematics.* Dietrich Reimer Verlag, 2001.

[182] J. Høyrup, *Lengths, widths, surfaces. A portrait of OB algrebra and its kin.* Springer, 2002.

[183] J. Høyrup, *Conceptual divergence –canons and taboos– and critique; reflections on explanatory categories.* Historia Mathematica **31**(2004), 129-147.

[184] F. Hultsch, *Heronis Alexandrini Geometricorum et Stereometricorum reliquiae,* Weidmann, 1864.

[185] A. Imhausen, *The Algorithmic Structure of the Egyptian Mathematical Problem Texts.* in: J.M. Steele & A. Imhausen (2002), 147-166.

[186] A. Imhausen, *Calculating the daily bread: Rations in theory and practice.* Historia Mathematica **30**(2003), 3-16.

[187] A. Imhausen, *Ägyptischen Algorithmen.* Harrasowitz Verlag, 2003b.

[188] D. Imhof, *Christoffel Plantin als uitgever van Simon Stevin.* in: Simon Stevin(2004), 43-48.

[189] C. Jacob, *La bibliothèque, la carte et le traité.* in: G. Argoud & J.Y. Guillaumin(1998), 19-37.

[190] S.A. Jayawardene, *Unpublished Documents Relating to Rafael Bombelli in the Archives of Bologna.* Isis **54**(1963), 391-395.

[191] S.A. Jayawardene, *Rafael Bombelli, Engineer-architect. Some unpublished Documents of the Apostolic Camera.* Isis **56**(1965), 298-306.

[192] S.A. Jayawardene, *The Influence of Practical Arithmetics on the Algebra (1572) of Rafael Bombelli.* Isis **64**(1973), 510-523.

[193] C.G. Jöcher, *Gelehrten Lexikon.* 11 vols. Olms, 1960-61 (reprint of 1750-1897).

[194] A.H.M. Jones, *Prosopography of the Later Roman Empire.* 3 vols. Cambridge UP, 1971.

[195] J.J. Joseph, *The Crest of the Peacock.* Penguin, 1992.

[196] P. Keyser, *Suetonius Nero 4.1.2 and the Date of Heron Mechanicus of Alexandria.* Classical Philology **83**(1988), 218-220.

[197] P. Kibre, *The Intellectual Interests reflected in Libraries of the Fourteenth and Fifteenth Centuries.* J. of the History of Ideas **7**(1946), 255-297.

[198] J. Klein, *Greek Mathematical Thought and the Origin of Algebra.* Dover Books, 1992.

[199] M. Kline, *Mathematical Thought from Ancient to Modern Times.* Oxford UP, 1972.

[200] E. Knobloch, *Studies in history of mathematics; dedicated to A.P. Youschkevitch.* Brepols, 2002.

[201] W.R. Knorr, *The Evolution of the Euclidean Elements.* D. Reidel, 1975.

[202] W.R. Knorr & G. Anawati, *Diophantus redivivus.* Archives internationales d'histoire des sciences **39**(1989), 345-357.

[203] W.R. Knorr, *Textual Studies in Ancient and Medieval Geometry.* Birkhäuser, 1989.

[204] W.R. Knorr, *Arithmêtikê stoicheîsis: On Diophantus and Hero of Alexandria.* Historia Mathematica **20**(1993), 180-192.

[205] W.R. Knorr, *Techniques of Fractions in Ancient Egypt and Greece.* in: J. Christianides(2004), 337-365.

[206] M. Kool, *Die conste vanden getale.* Verloren, 1999.

[207] N. Kruit & K.A. Worp, *Metrological notes on measures and containers of liquids in Graeco-Roman and Byzantine Egypt.* Archiv für Papyrusforschung und verwandte Gebiete **45:1**(1999), 96-127.

[208] L. Labowski, *Manuscripts from Bessarion's Library found in Milan. Bessarion Studies I.* Medieval and Renaissance Studies **5**(1961), 108-162.

[209] L. Labowski, *Bessarion's Library and the Biblioteca Marciana: six early inventories.* Storia e letteratura, 1979.

[210] S. Lampropoulou, *Hypatia, philosophe alexandrine.* Platon **29**(1977), 65-78. (in Greek)

[211] J. Lameer, *From Alexandria to Baghdad: Reflections on the Genesis of a Problematical Tradition.* in: G. Endress & R. Kok(1997), 181-191.

[212] M. Lang, *Numerical Notation on Greek Vases.* Hesperia **25**(1956), 1-24.

[213] P.L.M. Leone, *Maximi Monachi Planudes Epistolae.* Hakkert, 1991.

[214] M. Levey, *The Algebra of Abu Kamil.* University of Wisconsin Publications in Medieval Science, 1966.

[215] B. Levick, *Claudius.* Yale UP, 1990.

[216] N. Lewis, *The non-scholars of the Alexandrian Museum.* Mnemosyne **XVI** (1963), 257-261.

[217] H. L'Huillier, *Nicolas Chuquet; La Géométrie.* VRIN, 1979.

[218] E. Livrea, *A.P. 9.400: iscrizione funeraria di Ipazia.* Zeitschrift für Papyrologie und Epigraphik **117**(1997), 99-102.

[219] R. Lorch, *Greek-Arabic-Latin: the Transmission of Mathematical Texts in the Middle Ages.* Science in Context **14**(2001), 313-331.

[220] E. Lucas, *Récherches sur plusieurs ouvrages de Léonard de Pise.* Imprimerie des Sciences mathématiques et physiques, 1877.

[221] G. Luck, *Palladas, Christian or Pagan?* Harvard Studies in Classical Philology **62**(1958), 455-471.

[222] R.B. McClenon, *Leonardo of Pisa and his Liber Quadratorum.* in: F.J. Swetz(1994), 255-260.

[223] J.S. McKenzie, S. Gibson & A.T. Reyes, *Reconstructing the Serapeum in Alexandria from the Archeological Evidence.* Journal of Roman Studies **94**(2004), 73-127.

[224] R. MacLeod, *The Library of Alexandria.* Tauris, 2000.

[225] D.W. Maher & J.F. Makowski, *Literary evidence for Roman arithmetic with fractions.* Classical Philology **96**(2001), 376-399.

[226] A. Malet, *Renaissance notions of number and magnitude.* Historia Mathematica **33**(2006), 63-81.

[227] A. Malet, *Just before Viète: Numbers, polynomials, demonstrations and variables in Simon Stevin's* L'Arithmétique *(1585).* in: P. Radelet-de Grave (2008), 311-329.

[228] J. Mansfeld, *Prologomena mathematica: from Apollonius of Perga to late Neoplatonism.* Brill, 1998.

[229] Ammianus Marcellinus, *Res Gestae a fini Corneli Taciti.* (www.thelatinlibrary.com/ammianus.html)

[230] H.-I. Marrou, *Synesius of Cyrene and Alexandrian neoplatonism.* in: A.D. Momigliano, *The conflict between paganism and Christianity in the fourth century.* Clarendon, 1963.

[231] T. Martin, *Recherches sur la vie et les ouvrages d'Heron d'Alexandrie.* Mémoires présentés par divers savants à l'Académie des inscriptions et Belles-Lettres série I, tôme 4, 1854.

[232] A. Measson, *Alexandrea ad Aegyptum.* in : G. Argoud(1994), 9-52.

[233] D.J. Melville, *Weighing stones in ancient Mesopotamia*. Historia Mathematica **29**(2002), 1-12.

[234] D.J. Melville, *Poles and Walls in Mesopotamia and Egypt*. Historia Mathematica **31**(2004), 148-162.

[235] A. Meskens, *Wiskunde tussen Renaissance en Barok, aspecten van wiskunde-beoefening te Antwerpen 1550-1620*, Stadsbibliotheek Antwerpen, 1994a.

[236] A. Meskens, *Wine gauging at late sixteenth and early seventeenth century Antwerp*. Historia Mathematica **21**(1994b), 121-147.

[237] A. Meskens, *De prijs van de wetenschap*. De Gulden Passer **73**(1995), 83-106.

[238] A. Meskens, *De Thiende en andere Wisconstighe Ghedachtenissen*. in: V. Logghe (ed.), *Spiegheling en Daet; Simon Stevin van Brugghe (1548-1620)*. Bibliotheek De Biekorf, Brugge, 1996.

[239] A. Meskens, G. Bonte, J. De Groote, M. De Jonghe & D.A. King, *Wine-gauging at Damme. The evidence of a Late Medieval manuscript*. Histoire et mesure **XIV**(1999), 51-77.

[240] A. Meskens & N. van der Auwera, *Diophantos van Alexandrië; De zes boeken van de Arithmetika*. Hogeschool Antwerpen, 2006.

[241] A. Meskens, N. van der Auwera & P. Tytgat, *Aristarchos van Samos, Over de afstanden en de grootte van de zon en de maan*. Hogeschool Antwerpen, 2006b.

[242] A. Meskens, *Wiskunde tussen Renaissance en Barok, aspecten van wiskunde-beoefening te Antwerpen 1550-1620*. 2nd ed., Rondeel, 2009.

[243] R. Mett, *Regiomontanus in Italien*. Österreichische Akademie der Wissenschaften, 1989.

[244] M. Miller, *Bemerkungen zu Diophant von Pierre de Fermat*. Akademische Verlagsgesellschaft, 1932.

[245] L. Mohler, *Kardinal Bessarion als theologe, Humanist und Staatsmann*. 3 vols. F. Schöningh, 1923.

[246] J. Monfasani, *George of Trebizond; A Biography and a Study of his Rhetoric and Logic*. Brill, 1976.

[247] J. Monfasani, *Byzantine scholars in Renaissance Italy: Cardinal Bessarion and other emigrés*. Variorum, 1995.

[248] J-A Morse, *The Reception of Diophantus' "Arithmetic" in the Renaissance*. unpublished Ph.D., Princeton University, 1981.

[249] I. Müller (ed.), *Peri Ton Mathematon*, Apeiron **24/4**(1991).

[250] E. Müntz & P. Fabre, *La bibliothèque vaticane au XVe siècle d'après des documents inédits*. Bibliothèque des Écoles Française d' Athènes et de Rome **48**, 1887.

[251] K Muroi, *Small canal problems of Babylonian mathematics*. Historia Scientiarum(2) **1**(1992), 173-180.

[252] R. Netz, *The Shaping of Deduction in Greek Mathematics: a Study in Cognitive History*. Cambridge UP, 1999a.

[253] R. Netz, *Proclus' Division of the Mathematical Proposition into Parts: How and Why was it formulated?*. Classical Quarterly **49**(1999b), 282-303.

[254] R. Netz, *Greek Mathematicians, a Group Picture*. in: C.J. Tuplin & T.E. Rihll(2002a), 196-216.

[255] R. Netz, *Counter culture: towards a History of Greek Numeracy*. History of Science **40**(2002b), 319-332.

[256] R. Netz, *The Pythagoreans*. in: T. Koetsier & L. Bergmans, *Mathematics and the Divine: a Historical Study*. Elsevier, 2005, 77-98.

[257] O. Neugebauer, *Über eine Methode zur Distanzbestimmung Alexandria-Rom bei Heron*. Kongelige Danske Videnskabernes Selskabs Skrifter, **26/2**(1938), 21-24.

[258] V. Nutton, *Galen in Egypt*. in: J. Kollesch & D. Nickel, *Galen und das Hellenistische Erbe*. Franz Steiner Verlag, 1993, p. 11-31.

[259] J.A. Oaks & H.M. Alkhateeb, *Māl, enunciations and the prehistory of Arabic algebra*. Historia Mathematica **32**(2005), 400-425.

[260] J.A. Oaks & H.M. Alkhateeb, *Simplifying equations in Arabic algebra*. Historia Mathematica **34**(2007), 45-61.

[261] J.J. O'Connor & E.F. Robertson, *François Viète*. s.d. (www-groups.dcs.st-and.ac.uk).

[262] J.J. O'Connor & E.F. Robertson, *Abu Kamil Shuja ibn Aslam ibn Muhammad ibn Shuja*. 1999a (www-groups.dcs.st-and.ac.uk).

[263] J.J. O'Connor & E.F. Robertson, *Abu Jafar Muhammad ibn al-Hasan Al-Khazin*. 1999b (www-groups.dcs.st-and.ac.uk).

[264] J.J. O'Connor & E.F. Robertson, *Mohammad Abu'l-Wafa Al-Buzjani*. 1999c (www-groups.dcs.st-and.ac.uk).

[265] J.J. O'Connor & E.F. Robertson, *Hypatia of Alexandria*. 1999d (www-groups.dcs.st-and.ac.uk).

[266] J.J. O'Connor & E.F. Robertson, *Theon of Alexandria*. 1999e
(www-groups.dcs.st-and.ac.uk).

[267] J.J. O'Connor & E.F. Robertson, *Abu Ja'far Muhammad ibn Musa Al-Khwarizmi*. 1999f
(www-groups.dcs.st-and.ac.uk).

[268] J.J. O'Connor & E.F. Robertson, *Heron of Alexandria*. 1999g
(www-groups.dcs.st-and.ac.uk).

[269] J.J. O'Connor & E.F. Robertson, *Bernard Frenicle de Bessy*. 2000
(www-groups.dcs.st-and.ac.uk).

[270] R.A. Parker, *Demotic Mathematical Papyri*. Brown UP, 1972.

[271] R. Parkinson & S. Quirke, *Papyrus*. British Museum Press, 1995.

[272] E.A. Parsons, *The Alexandrian Library: Glory of the Hellenic World*. Elsevier, 1952.

[273] W.R. Paton, *The Greek Anthology 5*. Loeb-Heinemann, 1960.

[274] A. Pauly, G. Wissowa & W. Kroll, *Paulys Realencyclopädie der classischen Altertumswissenschaft neue Bearb. begonnen von Georg Wissowa; fortgeführt von Wilhelm Kroll*. Druckenmüller, 1958-

[275] R. Penella, *When was Hypatia born?*. Historia **33**(1984), 126-128.

[276] Philostorgius, E. Walford (trans.), *Epitome of the Ecclesiastical History of Philostorgius, compiled by Photius, Patriarch of Constantinople*. Henry G. Bohn, 1854. (www.tertullian.org/fathers)

[277] E. Picutti, *The Book of Squares of Leonardo of Pisa and the Problems of Indeterminate Analysis in palatine Codex 557*. Physis **21**(1979), 195-339.

[278] D. Pingree, *The Teaching of the 'Almagest' in Late Antiquity*. in: D. Barnes (ed.), *The Sciences in Greco-Roman Society*. Apeiron **27**(1994), 75-98.

[279] Plato, W.R.M. Lamb (trans.), *Plato in Twelve Volumes*. Vol. 3. Harvard UP and William Heinemann Ltd., 1925. (www.perseus.org)

[280] Plato, Harold N. Fowler (trans.), *Plato in Twelve Volumes*. Vol. 9. Harvard UP and William Heinemann Ltd., 1925. (www.perseus.org)

[281] Pliny the Elder, John Bostock (trans.), H.T. Riley (trans.), *The Natural History*. Taylor and Francis, 1855. (www.perseus.org)

[282] Pliny the Younger, *Panegyricus*.
(www.thelatinlibrary.com/pliny.panegyricus.html)

[283] Plutarch, Bernadotte Perrin (trans.), *Plutarch's Lives*. Harvard UP, 1920.
(www.perseus.org)

[284] D.T. Potts, *Before Alexandria, Libraries in the Ancient Near East.* in: R. MacLeod (2000), 19-34.

[285] Proclus, *Commentary on Euclid's Elements (excerpt of book I).* (www-history.mcs.st-and.ac.uk/Extras/Proclus_history_geometry.html)

[286] P. Radelet-de Grave, Liber Amicorum *Jean Dhombres.* Brepols, 2008.

[287] F.J. Ragep, S.P. Ragep (eds.), *Tradition, Transmission, Transformation.* Brill, 1996.

[288] S. Rappe, *The New Math: How to Add and to Subtract Pagan Elements in Christian Education.* in: Y.L. Too, *Education in Greek and Roman Antiquity.* Brill, 2001.

[289] R. Rashed, *Les travaux perdus de Diophante I.* Revue d'histoire des sciences **27**(1974a), 97-122.

[290] R. Rashed, *Résolution des équations numériques et algèbre: Saraf-al-Din-al Tusi, Viète.* Archives for the History of Exact Sciences **12**(1974b), 244-290.

[291] R. Rashed, *Les travaux perdus de Diophante II.* Revue d'histoire des sciences **28**(1975), 3-30.

[292] R. Rashed, *L'analyse diophantienne au Xe siècle: l'exemple de al-Khāzin.* Revue d'histoire des Sciences **32/3**(1979), 193-222.

[293] R. Rashed, *Diophante, Les Arithmétiques.* Tome III, Livres IV, Les Belles Lettres, 1984.

[294] R. Rashed, *Diophante, Les Arithmétiques.* Tome IV, Livres V, VI, VII, Les Belles Lettres, 1984.

[295] R. Rashed, *Notes sur la version arabe des trois premiers livres des Arithmétiques de Diophante et sur problème I.39.* Historia Scientiarum 4(1994), 39-46.

[296] R. Rashed, *The Development of Arabic Mathematics: between algebra and arithmetic.* D. Reidel, 1994b.

[297] R. Rashed, *Fibonacci et le prolongement Latin des mathématiques arabes.* Bolletino di Storia delle Scienze Matematiche **24**(2003), 53-73.

[298] K. Reich, *Diophant, Cardano, Bombelli, Viète: ein Vergleich ihrer Aufgaben* in: *Rechenpfenninge: Aufsätze zur Wissenschaftsgeschichte Kurt Vogel zum 80. Geburtstag.* Forschungsinstitut des deutschen Museums für die Geschichte der Naturwissenschaften und der Technik, 1968.

[299] K. Reich & H. Gericke, *François Viète: Einführung in die Neue Algebra.* Historiae scientiarum elementa **V**, 1973.

[300] L.D. Reynolds & N.G. Wilson, *Scribes and scholars: a guide to the transmission of Greek and Latin literature.* Clarendon, 1974.

[301] F.E. Robbins, *Mich.620: A Series of Arithmetical Problems*. Classical Philology **24**(1929), 321-329.

[302] F.E. Robbins, *Greco-Egyptian Arithmetical Problems Mich.4966*. ISIS **22**(1934), 95-103.

[303] G. Robins & C. Shute, *The Rhind Mathematical Papyrus; an Ancient Egyptian Text*. British Museum Press, 1987.

[304] E. Robson, *Mesopotamian Mathematics 2100-1600 B.C. Technical Constants in Bureaucracy and Education*. Clarendon, 1999.

[305] E. Robson, *Neither Sherlock Holmes nor Babylon: a reassessment of Plimpton 322*. Historia Mathematica **28**(2001), 167-206.

[306] E. Robson, *More than Metrology: Mathematics Education in an Old Babylonian Scribal School*. in: J.M. Steele & A. Imhausen(2002), 325-366.

[307] E. Robson, *Influence, ignorance, or indifference? Rethinking the relationship between Babylonian and Greek mathematics*. British Soc. for the History of Mathematics Bull. **4**(2005), 1-17.

[308] E. Robson, *Mesopotamian Mathematics: Some Historical Background*, in: V. Katz, *Using History to Teach Mathematics: An International perspective*. Math. Ass. of America, 2000, 149-158.

[309] A. Rome, *Commentaire de Pappus et Théon d'Alexandrie sur l'Almagest*. Biblioteca apostolica Vaticana, 1931-43.

[310] S. Rommevaux, A. Djebbar & B. Vitrac, *Remarques sur l'Histoire du Texte des Éléments d'Euclide*. Archive for the History of Exact Science **55**(2001), 221-295.

[311] D. Roques, *La famille d'Hypatie*. Révue d'études Grecques **108**(1995), 128-149.

[312] P.L. Rose, *Humanist Culture and Renaissance Mathematics: the Italian Libraries of the Quattrocento*. Studies in the Renaissance **20**(1973), 46-105.

[313] L. Russo, *The Forgotten Revolution*. Springer Verlag, 2004.

[314] K. Saito, *Reading ancient Greek mathematics*. in: E. Robson & J. Stedall, *The History of Mathematics*. Oxford UP, 2009, 801-826.

[315] D. Sakalis, *Die Datierung Herons von Alexandrien*. P.h.D. Dissertation, Universität zu Köln, 1972.

[316] N. Schappacher, *Diophantus of Alexandria: a Text and its History*. UFR de mathematiques et informatique, 2001. (schappa@math.u-strasbg.fr)

[317] W. Scheidel, *Death on the Nile*. Brill, 2001.

[318] F. Schmeidler, *Regiomontanus*. Opera Collectanea, 1972.

[319] W. Schmidt, *Heronis Alexandrini Opera quae supersunt omnia III*. B.G. Teubner, 1899.

[320] W.O. Schmidt, *Lateinische Literatur in Byzanz: die Übersetzungen des Maximos Planudes*. Jahrbuch des Österreichischen Byzantinischen Gesellschaft **17**(1968), 127-147.

[321] J. Schwartz, *La fin du Serapeum d'Alexandrie*. in: American Studies in Papyrology, 1966, 97-111.

[322] L. Séchan & Chatraine, *A. Bailly, Dictionnaire Grec-Français*. Hachette, 1950.

[323] Lucius Annaeus Seneca, John W. Basore (trans.), *Moral Essays*. 3 vols. Vol. II. The Loeb Classical Library, 1928-1935. (contains inter al. *De Tranquilitate Animi*). (www.stoics.com/seneca_essays_book_2.html)

[324] M. Serres (ed.), *A History of Scientific Thought*. Blackwell, 1995.

[325] J. Sesiano, *Les méthodes d'analyse indeterminée chez Abu Kamil*. Centaurus 21(1977), 89-105.

[326] J. Sesiano, *Die (arabish erhaltenen) Bücher IV-VII der* Arithmetika *des Diophantos aus Alexandria*. Arch. int. d'histoire des sciences **XXX**(1980), 183.

[327] J. Sesiano, *Books IV to VII of Diophantus' Arithmetica in the Arabic translation attributed to Qusta ibn Luqa*. Springer, 1982.

[328] J. Sesiano, *Sur une partie de l'Arithmetica de Diophante récemment retrouvé*. in: Les Sciences dans les textes antiques, Gent, 1986, 302-317.

[329] J. Sesiano, *The Appearance of Negative Solutions in Medieval Mathematics*. Archives for the History of Exact Sciences **32**(1995), 105-150.

[330] J. Sesiano, *Sur le Papyrus graecus genevensis 259*. Museum Helveticum **56**(1999), 26-32.

[331] J. Sesiano, *A Reconstruction of Greek Multiplication Tables for Integers*. in: Y. Dold-Samplonius, et al (2002), 45-56.

[332] I. Ševcenko, *Society and intellectual life in Late Byzantium*. Variorum, 1981.

[333] M.H. Shank, *The Classical Scientific Tradition in Fifteenth Century Vienna*. in: F.J. Ragep(1996), 115-136.

[334] N. Sidoli, *Heron's* Dioptra *35 and Analemma methods: An Astronomical Determination of the Distance between Two Cities*. Centaurus **47**(2005), 236-258.

[335] L.E. Sigler, *Leonardo Pisano Fibonacci, The book of squares*. Academic Press, 1987.

[336] S. Singh, *Fermat's Last Theorem*. Fourth Estate, 1997.

[337] J.G. Smyly, *The Employment of the Alphabet in Greek Logistic.* in: *Mélanges Nicole*, W. Kündig, 1905, 515-530.

[338] Socrates Scholasticus, A.C. Zenos (trans.), *The Ecllesiastical History of Socrates Scholasticus, Revised, with Notes*. Wm. B. Eerdmans Publ. Co, 1886. (www.ccel.org/ccel/schaff/npnf202.pdf)

[339] Aelius Spartianus, David Magie (trans.), *The Life of Antoninus Caracalla.* The Loeb Classical Library, 1924.
(members.aol.com/heliogabby/bio/caracall.htm)

[340] O. Spengler, *Der Untergang des Abendlandes: Umrisse einer Morphologie der Weltgeschichte.* Beck, 1923.

[341] Strabo, H.C. Hamilton (ed.), W. Falconer (ed.), *Geography. The Geography of Strabo. Literally translated, with notes, in three volumes.* George Bell & Sons, 1903. (www.perseus.org)

[342] K. Staikos, T. Cullen, *The History of the Western Library.* Oak Knoll Press, HES & DeGraaf, 2004.

[343] E.S. Stamatis, *Diophantus als Mathematiker*. Das Altertum **19**(1973), 156-164.

[344] E.S. Stamatis, *Rekonstruktion des griechischen Textes des fehlenden Beweises der Aufgabe V19 des Diophantos von Alexandrien.* Miscellanea Critica Teubner **I**(1964), 265-267.

[345] J.M. Steele & A. Imhausen, *Under One Sky: Astronomy and Mathematics in the Ancient Near East.* Ugarit-Verlag, 2002.

[346] S. Stevin, *L'Arithmétique*. Plantin, 1585.

[347] S. Stevin, *L'Arithmétique*. Elzevier, 1625.

[348] C.L. Stinger, *The Renaissance in Rome.* Indiana UP, 1985.

[349] J. Straub, *Severus Alexander und die Mathematici.* in: G. Alföldy, *Bonner Historia-Augusta Colloquium 1968/69*. Habelt, 1970.

[350] D.J. Struik, *Minoan and Mycenenan Numerals*. Historia Mathematica **9**(1982), 54-58.

[351] V.V. Struve & B. Turaev, *Mathematischer Papyrus des Staatlichen Museums der Schönen Künste in Moskau. Quellen und Studien zur Geschichte der Mathematik; Abteilung A: Quellen 1.* Springer, 1930.

[352] C. Tranquillus Suetonius, Maximilian Ihm (ed.), J. C. Rolfe (trans.), *The lives of twelve Caesars, De Vita Caesarum–Divus Claudius.* The Loeb Classical Library, 1913-1914.
(www.perseus.org and penelope.uchicago.edu/Thayer/E/Roman/home)

[353] H. Suter, *Die Mathematiker und Astronomen der Araber und ihre Werke*. B.G. Teubner, 1900.

[354] H. Suter, *Nachtrage und Berichtigungen zu die Mathematiker und Astronomen der Araber und Ihre Werke*. B.G. Teubner, 1902.

[355] F.J. Swetz (ed.), *From Five Fingers to Infinity*. Open Court, 1994.

[356] J.D. Swift, *Diophantus of Alexandria*. The American Math. Monthly **63** (1956), 163-170 (reprint in: F.J. Swetz(1994), 182-188).

[357] A. Szabó, *The beginnings of Greek mathematics*. D. Reidel, 1978.

[358] Cornelius Tacitus, Alfred John Church (ed.), William Jackson Brodribb (ed.), Sara Bryant (ed. for Perseus), *The Histories*. Random House, Inc., 1873, repr. 1942. (www.perseus.org)

[359] P. Tannery & C. Henry, *Œuvres de Fermat I: œuvres mathématiques et observations sur Diophante*. Gauthiers-Villars, 1891.

[360] P. Tannery, *Diophanti Alexandrini, Opera Omnia cum Graecis Commentariis*. 2 vols. B.G. Teubner, 1895.

[361] P. Tannery, J.L. Heiberg, H.G. Zeuthen (ed.), *Mémoires scientifiques*. 17 vols. Privat, 1912-1940.

[362] P. Tannery, *Quadrivium de Georges Pachymère*. Studi e testi **94**, 1940.

[363] P. Tannery, *Diophanti Alexandrini, Opera Omnia cum Graecis Commentariis*. 2 vols. B.G. Teubner, 1974. (reprint of 1895)

[364] P. Tannery, *Psellus sur Diophante*, Mathematik und Physik, s.d.

[365] L. Taran, *Commentary to Nicomachus' Introduction to Algebra*. American Philosophical Society, 1969.

[366] R. Taylor & A. Wiles, *Ring-theoretic properties of certain Hecke algebras*. Annals of Mathematics **141**(1995), 553-572.

[367] Tertullian, Alexander Souter (trans.), *Tertullian's defence of the Christians against the heathen*. (www.tertullian.org)

[368] Y. Thomaidis, *A Framework for defining the Generality of Diophantos' Methods in "Arithmetica"*. Archive for History of Exact Science **59**(2005), 591-640.

[369] L. Thorndike, *Giovanni Bianchini in Paris Manuscripts*. Scripta Mathematica **16**(1950), 5-12 & 169-180.

[370] H. Thurston, *Greek mathematical astronomy reconsidered*. ISIS **93**(2002), 58-69.

[371] M.N. Tod, *Three Greek Numeral Systems*. J. of Hellenic Studies 33(1913), 27-34.

[372] M.N. Tod, J. Breslin, *Ancient Greek numerical systems: six studies.* Ares, 1979.

[373] G.J. Toomer, *Lost Greek Mathematical Works in Arabic translation.* in: J. Christianides(2004), 275-284.

[374] J. Tropfke, *Geschichte der Elementarmathematik.* 4. Aufl. W. De Gruyter, 1980.

[375] C.J. Tuplin & T.E. Rihll, *Science and Mathematics in Ancient Greek Culture.* Oxford UP, 2002.

[376] Johannes Tzestzes, *Prologomena de Comoedia.* (www.attalus.org/translate/poets.html#Lycophron2)

[377] G. Vanden Berghe, *Simon Stevin, een leven in de schaduw.* in: *Simon Stevin*(2004), 19-24.

[378] N. van der Auwera & A. Meskens, *Apicius, De Re Coquinaria; De Romeinse kookkunst.* Koninklijke Bibliotheek, 2001.

[379] W. Van Egmond, *The Commercial Revolution and the Beginnings of Western Mathematics in Renaissance Florence 1300-1500.* Unpublished Ph.D. Indiana University, 1976.

[380] W. Van Egmond, *A catalog of François Viète's printed and manuscript works.* in: M. Folkerts, U. Lindgren(1985), 359-393.

[381] H. Van Staden, *Galen's Alexandria.* in: W.V. Harris & G. Ruffini, *Ancient Alexandria between Egypt and Greece.* Brill, 2004, 179-185.

[382] J.J. Verdonck, *Petrus Ramus en de wiskunde.* Van Gorcum, 1966.

[383] P. Ver Eecke, *Diophante d'Alexandrie, Les six Livres arithmétiques et le livre des nombres polygones.* Desclée-De Brouwer, 1926.

[384] P. Ver Eecke, *Léonard de Pise, Le livre des nombres carrés.* Desclée-De Brouwer, 1952.

[385] F. Viète-F. van Schooten, *Francisci Viètæ Opera mathematica: in unum volumen congesta, ac recognita / operâ atque studio Francisci à Schooten.* Elzevier, 1646.

[386] B. Vitrac, *Euclide et Héron: deux approches de l'enseignements des mathématiques dans l'Antiquité?.* in: G. Argoud(1994), 121-146.

[387] B. Vitrac, *Promenade dans les préfaces des textes mathématiques grecs anciens.* in: P. Radelet-de Grave(2008), 519-556.

[388] K. Vogel, *Die algebraischen Probleme des Mich.620.* Classical Philology **25** (1930), 373-375.

[389] K. Vogel, *Zur Berechnung der quadratischen Gleichungen bei den Babyloniern.* in: J. Christianides(2004), 265-273.

[390] Vopiscus, *Historia Augusta, The Lives of Firmus, Saturninus, Proculus and Bonosus.* The Loeb Classical Library, 1932.
(penelope.uchicago.edu/Thayer/E/Roman/home)

[391] D. Wagner, *Zur Biographie des Nicasius Ellebodius (+1577) und seinen "Notæ" zu den aristotelischen Magna Moralia.* Carl Winter Universitätsverlag, 1973.

[392] E. Walser, *Poggius Florentinus, Leben und Werke.* G. Olms Verlag, 1974.

[393] H.-J. Waschkies, *Anfänge der Arithmetik im Alten Orient und bei den Griechen.* Verlag B.R. Grüner, 1989.

[394] A. Wasserstein, *Thaetetus and the History of the Theory of Numbers.* Classical Quarterly **8**(1958), 165-179.

[395] W.C. Waterhouse, *Harmonic Means and Diophantus.* Historia Mathematica **20**(1993), 89-91.

[396] E.J. Watts, *City and School in late antique Athens and Alexandria.* University of California Press, 2006.

[397] A. Weil, *Number Theory. An Approach through History from Hammyurapi to Legendre.* Birkhäuser, 1984.

[398] R. Weiss, *The Renaissance Rediscovery of Classical Antiquity.* Blackwell, 1973.

[399] C. Wendel, *Planudea.* Byzantinisches Zeitschrift **40**(1940), 406-445.

[400] C. Wendel, *Planudes.* Metzler, 1942.

[401] J.L. Wetherell, *Bounding the Number of Rational Points on Certain Curves of High Rank.* unpublished Ph.D. dissertation, University of California at Berkeley, 1998.

[402] F. Wiering, *Bernardino Baldi, Cronica de matematici (1707): a machine-readable edition.*
(euromusicology.cs.uu.nl:6334/dynaweb/info/persinfo/balcro/)

[403] A. Wiles, *Modular elliptic Curves and Fermat's Last Theorem.* Annals of Mathematics **141**(1995), 443-551.

[404] B. Williams & R.S. Williams, *Finger Numbers in the Graeco-Roman World and the Early Middle Ages.* ISIS **86**(1995), 587-608.

[405] A.M. Wilson, *The Infinite in the Finite.* Oxford UP, 1995.

[406] N. Wilson, *Scholars of Byzantium.* Duckworth, 1983.

[407] A. Winger-Trennhaus, *Regiomontanus als Frühdrucker in Nürnberg.* Verein für Geschichte der Stadt Nürnberg, 1991.

[408] M. Wood, *In the Footsteps of Alexander the Great,* University of California Press, 1997.

[409] G. Xylander, *Diophanti Alexandrini Rerum Arithmeticarum libri sex.* E. Episcopium, 1575.

[410] F. Yates, *The French Academies of the Sixteenth Century.* Warburg Institute, 1947.

[411] E. Zinner, *Regiomontanus.* Zeller, 1968.

Index